RISK ANALYSIS IN FOREST MANAGEMENT

Managing Forest Ecosystems

Volume 2

Series Editors:

Klaus von Gadow

Georg-August-University,
Göttingen, Germany

Timo Pukkala

University of Joensuu,
Joensuu, Finland

and

Margarida Tomé

Instituto Superior de Agronomía,
Lisbon, Portugal

Aims & Scope:

Well-managed forests and woodlands are a renewable resource, producing essential raw material with minimum waste and energy use. Rich in habitat and species diversity, forests may contribute to increased ecosystem stability. They can absorb the effects of unwanted deposition and other disturbances and protect neighbouring ecosystems by maintaining stable nutrient and energy cycles and by preventing soil degradation and erosion. They provide much-needed recreation and their continued existence contributes to stabilizing rural communities.

Forests are managed for timber production and species, habitat and process conservation. A subtle shift from *multiple-use management* to *ecosystems management* is being observed and the new ecological perspective of *multi-functional forest management* is based on the principles of ecosystem diversity, stability and elasticity, and the dynamic equilibrium of primary and secondary production.

Making full use of new technology is one of the challenges facing forest management today. Resource information must be obtained with a limited budget. This requires better timing of resource assessment activities and improved use of multiple data sources. Sound ecosystems management, like any other management activity, relies on effective forecasting and operational control.

The aim of the book series ***Managing Forest Ecosystems*** is to present state-of-the-art research results relating to the practice of forest management. Contributions are solicited from prominent authors. Each reference book, monograph or proceedings volume will be focused to deal with a specific context. Typical issues of the series are: resource assessment techniques, evaluating sustainability for even-aged and uneven-aged forests, multi-objective management, predicting forest development, optimizing forest management, biodiversity management and monitoring, risk assessment and economic analysis.

The titles published in this series are listed at the end of this volume.

Risk Analysis in Forest Management

Edited by

Klaus von Gadow

Georg-August-University, Göttingen, Germany

KLUWER ACADEMIC PUBLISHERS

DORDRECHT / BOSTON / LONDON

A C.I.P. Catalogue record for this book is available from the Library of Congress.

ISBN 978-90-481-5683-2

Published by Kluwer Academic Publishers,
P.O. Box 17, 3300 AA Dordrecht, The Netherlands.

Sold and distributed in North, Central and South America
by Kluwer Academic Publishers,
101 Philip Drive, Norwell, MA 02061, U.S.A.

In all other countries, sold and distributed
by Kluwer Academic Publishers,
P.O. Box 322, 3300 AH Dordrecht, The Netherlands.

Printed on acid-free paper

Foreword

Due to the long-term planning horizons and the great variety of natural, economic and operational hazards affecting forest ecosystems, uncertainty and multiple risk are typical aspects of forest management. Applications of risk analysis are surprisingly rare, in spite of the rich assortment of sophisticated forest planning tools that are available today.

Damage is loss expressed in monetary terms. The *damage potential* includes all the potential threats within a given *hazard domain*. *Risk* has been defined as the expected loss due to a particular hazard for a given area and reference period. Disregarding the cost of capital, the *expected loss* (*r*) may be calculated as the product of the *damage* (*s*) expressed in monetary terms and its *probability* (*p*), $r = s \cdot p$. Example: The probability of a spruce tree being damaged by wind increases with increasing tree size, whereas the damage itself is the result of an increase in the harvesting costs and a decrease in the log price. *Risk assessment* is a formal procedure for quantifying risk with regard to the damage potential, including all the possible threats within a given hazard domain. Thus, risk is not the same as uncertainty. Uncertainty presents a risk if the result of the uncertainty is an expected loss. It is important to delineate the system and to identify potential hazard factors. More difficult is the assessment of hazard factor probabilities because forest management, in contrast to industrial safety management, is affected by exogeneous hazards that cannot be controlled. A practical approach which has been applied in forest planning models is to estimate age-dependent cumulative survival rates for a given set of hazard factors.

The subject of a risk analysis is a given system which includes valuable objects that might be lost or damaged. The system needs to be defined in terms of time and space and the seriousness of the expected hazard. Within the scope of forest planning, the scales of time and space may range from a year to a century and from an individual tree to an entire forest region. Short- to medium-term tree-based risk analysis is essential for economic evaluation of removal decisions in selective thinning models. Medium-term stand-based risk analysis is concerned with specific abiotic or biotic hazards affecting particular types of stand structure, species or age class. A typical problem in forest scenario planning is to predict major losses on a compartment or forest level. Long-term risk analysis for forest regions is concerned with the choice of tree species and silvicultural systems.

The objective of this particular volume within the book series *Managing Forest Ecosystems* is to present state-of-the-art research results, concepts and techniques regarding the assessment and evaluation of natural hazards and the analysis of risk and uncertainty relating to forest management. Various aspects of risk analysis are covered, including examples of specific modelling tools. The book is divided into three sections covering ecological perspectives, applications in engineering and planning and methods applicable to economics and policy.

The first section presents two contributions covering forest ecosystem and conservation issues. The contribution by *Annika and Jyrki Kangas* from Finland analyses of the impacts of forestry on wildlife populations. Population viability is estimated as a function of forest-based variables, initial population size and environmental change, using a stochastic population dynamics model. The subject of the second paper by *Michael Bredemeier, Norbert Lamersdorf, Hubert Schulte-Bisping and Burckhardt von Lüpke* is an evaluation of environmental changes and the associated modification of site conditions and ecosystem transformations, including CO_2 enrichment and enhanced nitrogen deposition, soil acidification and soil compaction.

The second section is concerned with applications in engineering and planning. Artificial intelligence applications in risk analysis are presented by *Daniel Schmoldt* who examines knowledge-based systems, fuzzy logic, artificial neural networks and Bayesian belief networks and their application to risk analysis in the context of forest ecosystem management. This is followed by *Kurt Hollenstein's* multi-level risk analysis approach dealing with land use decisions in Switzerland and including risk variables such as loss of life, loss of ecosystem functionality, greenhouse gas emission and financial losses. *Atsushi Yoshimoto* presents sophisticated quantitative applications of risk analysis, including stochastic dynamic programming and estimation of nonlinear models to stochastic phenomena, in the context of timber harvest planning.

In the third section the emphasis is on economics and policy. *David Klemperer* gives an overview of risk in the context of financial analysis, including present value probability histograms, simulation, problems with the weighted average cost of capital and the capital asset pricing model. This is followed by an empirical case study presented by *Gerhard Weiss* who is able to show differences in the perception of and attitude towards risk indifferent public administrations. The risk policies of the Austrian forestry administration and the torrent and avalanche control service are evaluated and compared. Finally, based on economic theory and optimization, *Martin Moog* uses examples of serious hazards, such as storm damage, to evaluate forest management decision alternatives and cutting strategies in spruce stands in Germany.

A potential hazard presents a risk if it occurs with a probability greater than zero and if its occurrence will cause damage to a valuable object. Although there is evidence that extensive experience about various kinds of damage to forest trees exists, models for predicting potential hazards are rarely applied in practice. This volume contains a wealth of methods, concepts and specific case studies with worked examples dealing with hazard and risk analysis, with contributions from the natural, social and engineering sciences. If technical knowledge is a constraint to the implementation of risk analysis in forest management, this book contains a wealth of information which may be useful to company management, the legal and policy environment, forestry administrators and students.

The valuable contributions by our anonymous referees in the form of rigorous criticism and positive suggestions is gratefully acknowledged. The outstanding support and encouragement of Kluwer's publishing editor Ursula Hertling is appreciated. Finally, I wish to thank Sonja Rüdiger and Marga von Gadow for their diligent proofreading and editing assistance.

Klaus v. Gadow, September 2000

Contents

Ecosystem Perspectives

Engineering and Planning

Economics and policy

Section 1

Ecosystem perspectives

FOREST MANAGEMENT PLANNING FOR MAINTAINING THE VIABILITY OF WILDLIFE POPULATIONS

Annika S. Kangas[1] & Jyrki Kangas[2]*

ABSTRACT

Ecological information about the impacts of forestry on wildlife populations have rarely been used in calculations of forest management planning. Some applications exist, where models predicting sizes of some wildlife populations have been used in optimisation calculations. When such models are used, it is important to take the inherent uncertainty into account. It is essential to know the probability that the size of the population of interest would fall below presumably critical limits. One useful tool in decision-making concerning wildlife populations is risk analysis. In risk analysis, viability of the populations is assessed by a stochastic population dynamics model. The population viability may, for instance, be expressed as the probability of the population surviving the next 1000 years. To make use of risk analysis in forest management planning, viability needs to be expressed as a function of forest-based characters. It may also be necessary to sort out the risk due to different factors, such as initial population size and changes in the environment. In this paper, the possibility of using information about wildlife populations in forest management planning is considered. Special attention is given to the handling of uncertainty inherent in most ecological information.

Keywords: *fuzzy decision making, optimisation, population dynamics modelling, viability analysis, uncertainty*

[1] Researcher, University of Joensuu, Finland

[2] Station Head, Kannus Research Station of the Finnish Forest Research Institute, Finland

[*] The work was carried out in the project "Integrating landscape ecology into forest planning" that belongs to the Finnish Biodiversity Research Programme FIBRE, funded by the Finnish Academy. The work was also funded by the Finnish Forest Research Institute, the Kannus Research Station, and the Ministry of Agriculture and Forestry of Finland.

K. von Gadow (ed.), Risk Analysis in Forest Management, 1–20.

1 Introduction

In forest management planning the target is to find the treatment schedule which best meets the objectives set for the forests. In recent years, ecological aspects have been given more and more attention in forest management planning. From the ecological perspective, one key target in planning is to maintain the viability of the species persisting in the planning area. A viable population could be defined, for instance, as a population which has a 99% probability of surviving the next 1000 years (Shaffer, 1981).

The most important reason for not using information about wildlife species in calculations of forest management planning has been the lack of information that can be used for numerical planning models. These models, usually based on mathematical optimisation, require ecological knowledge to be presented in relatively simple forms. Also, the number of different species is usually too high to consider all species in planning. On some species there may not be any information available. Thus, planning needs to focus on the most vulnerable species, species groups or on species that represent the reactions of other species (so-called key species).

All available information should be utilised in decision-making. Information available on species of interest may include experts' judgements, models and empirical data. The most apparent type of information concerning a wildlife population is a model describing population size as a function of some characteristics of the planning area (e.g. Burgman et al. 1994, Haight & Travis, 1997). Information about habitat requirements of the species is, however, more common (e.g. Taylor & Haseler, 1993). It may be easier to gather information about the habitat requirements than about the species themselves. Making inferences about the species from habitat information, may however, be more difficult than it seems (e.g. McCarthy et al., 1994). Expert judgement can be used, for example, to assess the relative benefits of a plan from the standpoint of wildlife populations (e.g. Kangas et al., 1993, Alho et al., 1996). Information obtained from risk (or viability) analysis concerning wildlife populations is increasingly used in decision-making. In risk analysis, the probability of the population becoming extinct (or declining to a critical level) under predefined conditions is assessed (e.g. Burgman et al., 1993). Risk analysis is based on a stochastic model of population dynamics.

Uncertainty is often ignored in forest management planning calculations. In ecological considerations, however, the uncertainty may be considerable. Consequently, information on the uncertainty involved is essential in environmental decision-making. One refers to decision-making under risk, if the decision alternatives with their possible consequences, and the probabilities of these consequences occurring, are known. If the probabilities or even the consequences of the alternative actions are not known, the decision is made under uncertainty.

Different decision-makers can take different attitudes towards risk and uncertainty, which may lead to different priority orders among choice alternatives. Maximisation of expected utility is a common decision strategy when supporting risky choices. When dealing with ecological objectives, decisions are, in principle, always made under uncertainty. However, it is usually possible to estimate the probabilities associated to consequences of choice alternatives and to use these in decision making. This being the case, risk management methods can be applied also to support decision-making under uncertainty.

For decision support under uncertainty different strategies have been developed. For example, according to the Maximin-criterion, the alternatives are ranked according to the worst possible consequences (risk avoiding behaviour). On the other hand, according to the Maximax-criterion, the alternatives are ranked according to the best possible consequences (risk taking behaviour). The Hurwicz-criterion is a combination of these two strict rules; the alternative with the greatest weighted mean of the worst and the best possible outcomes is chosen. Here, the weights reflect the attitude towards risk; e.g., for a risk neutral decision-maker the weights are equal. A more general criterion, which produces the above mentioned criteria as special cases, has been presented by Kangas (1992, 1994). In the well-known von Neumann-Morgenstern utility function, risk attitudes are dealt with implicitly; the form of the utility function describes the decision-maker's attitude towards risk (von Neumann & Morgenstern, 1947).

In terms of environmental impacts, the consequences of decisions may be irreversible. Therefore, it has been often required that viability of populations should be given priority over monetary goals. This means implementing a precautionary principle (Faucheux & Froger, 1995). When one decision is irreversible (cutting an old-growth forest) and the other is reversible (conserving an old-growth forest), there is a value associated with the reversible decision, a so-called quasi-option value (Reed, 1993).

2 Ecological information

2.1. Population models and risk analysis

A simple model describing the development of the population can be presented as (e.g. Burgman et al., 1993)

$$N_{t+1} = N_t + N_t b - N_t d = N_t (1+r). \tag{1}$$

In this model, N_t is the size of the population at time t, b is the birth rate and d is the death rate. Thus, r is the growth rate (discrete time). This model predicts the development of the population in a deterministic manner. If birth rate is greater than death rate, the population grows indefinitely; otherwise the population eventually becomes extinct. Often, for reasons of simplicity, only the females are considered in the model.

In reality, the growth of a population is not deterministic; which means there is stochastic variation. For example, the number of offspring varies among females. This demographic variation can be assumed to follow, for example, a binomial distribution. Thus, the number of individuals that die or are born can be described as (e.g. Kokko et al., 1997, Drechsler et al., 1998)

$$N_{t+1}=N_{t,survived}+N_{t,born}, \tag{2}$$

where $N_{t,born} \sim BIN(N_t,b)$ and

$N_{t,\,survived} \sim BIN(N_t,(1-d))$.

With these assumptions, it is possible to calculate the probability of the population becoming extinct. The probability can be calculated, for example, by stochastically simulating a large number of realisations of the development of the population. The realisations are obtained by sampling from the binomial distribution. Then, the number of realisations in which the size of the population goes to zero is counted. This number, in proprtion to the number of realizations, is an assessment of the risk the population is facing due to demographic variation.

In general, the effect of demographic variation is relatively small, if the population is large (Lande, 1993). Variation at population level, such as variation due to weather conditions, may be much more important. This kind of variation (e.g. annual or seasonal) can be described by including random variation into the birth and death rates. The variation is usually assumed to follow some probability distribution. With this information, it is possible to calculate the risk the population is facing due to variation in the birth and death rates in a similar fashion as above. The growth rate of the population may also depend on the density of the population. One example of such a model is the logistic growth model

$$N_{t+1} = N_t\left[1 + r_t\left(1 - \frac{N_t}{K}\right)\right], \tag{3}$$

where K is the carrying capacity of the environment (e.g. Burgman et al., 1993).

The probability of dying or giving birth is not equal among individuals. It can be assumed to depend on age or development stage of individuals. The age structure of the population can be taken into account by using a Leslie matrix approach (Leslie, 1945). In addition to several age-groups, the population may consist of several sub-populations. Combination of the sub-populations is called a metapopulation (Hanski, 1989, 1991; Hanski & Gilpin, 1991). In a metapopulation the individuals may migrate from one sub-population to another. The development of one sub-population may be described with (e.g. Burgman et al., 1993)

$$N_{i(t+1)} = N_{it}(1 + r_i) - \sum_{j=1}^{n} Eij + \sum_{j=1}^{n} Eji, \tag{4}$$

where E_{ij} describes the migration from sub-population i to j. In a metapopulation model, a spatial level of variation can be included. For example, the growth rates may vary among the sub-populations. This variation in the vital rates may be positively correlated among the sub-populations. The stronger this correlation, the larger the risks concerning the whole metapopulation (Harrison & Quinn, 1989; Burgman et al., 1993). On the other hand, strong migration will reduce the extinction risk for the metapopulation, by allowing a recovery of an extinct sub-population. Another possible factor is the risk of catastrophic events such as forest fires (e.g. Lande, 1993).

Recently, several studies assessing the viability of endangered or vulnerable species through risk analysis have been carried out. Risk analysis can be used to identify what species are at greatest risk, what critical factors affect the risks and what can be done to reduce these risks. Risk assessments have been made, for example, for spotted owl (*Strix occidentalis*) (Lamberson et al., 1992), helmeted honeyeater (*Lichenostomus melanops cassidix*) (McCarthy, 1996), seal (*Halichoerus grypus* and *Phoca hispida)* (Kokko et al.,1997), Hawaiian stilt (*Himantopus mexicanus knudseni*) (Reed et al., 1997), orange-bellied parrot (Neophema Chrysogaster) (Drechsler et al., 1998), great tit (*Parus major*) (Sæther et al., 1998) and Capercaillie (*Tetrao urogallus*) (Kangas & Kurki, 2000).

The risk assessments considered above produce information about the performance of the population under stable conditions. However, risk analysis can also be used to explore the consequences of different management options (e.g. Drechsler et al.,1998). For example, the feasibility of reintroduction of Capercaillie to Southern Scotland (Marshall & Edward-Jones, 1998) has been assessed. Also, the best conservation strategy has been determined, i.e. the size and distribution of the conservation areas (Salwasser et al., 1984, Akçakaya et al., 1995). The effects of different hunting strategies have also been described using risk analysis (Kokko et al., 1997, Kokko & Lindström, 1998).

The risk assessments are useful in forest management planning only if the effect of forest management on the risk can be assessed. This is possible, if one or more of the vital parameters depend on the characteristics of the forests in the area. Otherwise, all management options are assumed equally good from the population's point of view. A further complication is that the planning area usually cannot be divided to "suitable" and "nonsuitable" areas, but each different developmental stage of forests is a more or less suitable habitat for the species of interest. For example, for northern spotted owl, the risk faced by the population has been assessed as a function of the proportion of old growth forests (McKelvey, 1996). Furthermore, distribution of the suitable habitat patches in the area may have a profound effect on viability: the sufficient percentage of suitable habitat may depend on the spatial connectivity of these patches (e.g. Swart & Lawes 1996).

Often, risk assessments produce a sharp threshold value, i.e. a value of forest characteristics, below which the viability quickly decreases. Such threshold values can be used directly in the decision support. However, the precise location of the threshold has been found to be quite sensitive to the parameters of the model (McKelvey, 1996). The location of the threshold also depends on the time horizon considered: the longer the time horizon, the more restrictive the threshold value becomes.

One indicator, such as the probability of survival for 100 years, may, however, be misleading. For example, the choice of the time horizon considered may markedly alter the risk analysis results (McKelvey, 1996). If the time range considered is long enough, any population becomes extinct under continuing risk. Another common indicator of ecological risk is the expected time to extinction. The distribution of the extinction time is usually very skewed. A few realisations with a very long time to extinction may greatly increase the expected extinction time, so that the extinction time in most of the realisations is shorter than the expected time (Burgman et al., 1993). Thus, a set of indicators giving information on several aspects of risk is more useful in decision making. For example, one possibility is to examine the hazard profile. The τ-year hazard is the probability that a population, survived to time t, will become extinct within the next τ years. The hazard profile, i.e. the hazards at different times, can then be used to sort the short-term risks that are due to the initial size of the population from persistent risks due to habitat quality (McKelvey, 1996).

The values of the parameters used in the analysis are usually quite uncertain. Thus, sensitivity analysis for the parameters is essential (e.g. McCarthy et al., 1995, 1996; Drechsler et al., 1998). There may also be uncertainty about the shape of the models used to describe population development. Different assumptions, for example, about density dependence, may lead to large differences in risk analysis results (Ginzburg et al., 1990; Mills et al., 1996; Sæther et al., 1998; Kangas & Kurki, 2000). It has been suggested that risk analysis should be carried out with several alternative model assumptions. An (weighted) average from several models (e.g. Raftery et al., 1995, Draper, 1995) could then be utilised to assess risk. In many cases, however, it may be sufficient to assess an upper bound for the risk. The upper bound can in most cases be obtained by using a density-independent model (Ginzburg et al., 1990).

If there are no data available about the stochasticity involved, the usefulness of the results can be called into question (Caughley, 1994; Sæther et al., 1998). McKelvey (1996) states that the true significance of a computed index value in a real population is unknown, and is also unknowable. The risk analysis, however, can provide substantial insight to the population dynamics and the effects of management strategies.

2.2. Expert judgements

In many cases, available empirical information is not detailed or structured enough to form a simulation model describing the population dynamics or to perform risk analyses. In such a case it is possible to use expert judgement to assess the missing parameters. The expert judgement can be utilised using Bayesian population dynamics modelling for formulating prior distributions for the parameters (Raftery et al., 1995; Kangas & Kurki, 2000). If most of the required parameters are missing, it is possible to apply expert judgement to assess the critical values of the habitat. Also in this case, available empirical data on the occurrence of the species could be used to support expert judgements about such threshold values. Another possibility is to holistically rank the plans from a population's standpoint using expert opinions (e.g. Kangas et al., 1993).

If the information available for planning is based on pure expert judgements, the uncertainty is difficult to estimate, but especially important to take into account. The uncertainty in expert judgements can be analysed on the basis of the internal consistency of the expert's opinions, or the differences within a group of experts (Alho et al., 1996; Alho & Kangas, 1997). The expert may also be asked to assess how certain he/she is about his/her judgements (Leskinen & Kangas, 1998). These sources of uncertainty might not, however, cover all the uncertainty in the judgements.

3 Ecological knowledge in forest management planning

3.1. Linear Programming method

Linear Programming (LP) is a method commonly used in forest management planning. Using LP, the planning problem is described with an objective function which is to be maximised

$$\max \sum_{j=1}^{n} c_j x_j \tag{5}$$

subject to constraints

$$\sum_{j=1}^{n} a_{ij} x_j \geq b_i,\ i = 1,\dots,p \tag{6}$$

In this formulation, x_j is the area treated with alternative j, c_j is the coefficient that represents how much the area contributes to the objective variable, a_{ij} is a coefficient representing how much the area treated with treatment j produces or uses commodity i, and b_i is a coefficient which represents how much of commodity i is required, or can be utilized.

The population models can be used in a LP-problem, for example, by taking the size of the wildlife population as an objective or as a constraint. Some forest management planning models include components for dealing with wildlife populations. In one such model, FORPLAN, the population size of a few species can be taken as a decision criterion (Burgman et al., 1994). However, the sizes of animal populations alone may not be the major concern. In addition, the decision-maker may want to keep the populations viable although he/she manages the forests for other purposes, such as wood production or recreation. In such case, the risk faced by the population can be used as an objective or as a constraint in planning.

The usual LP method does not consider the underlying uncertainty. Here, it is conceivable that the obtained solutions may not be feasible. The decision-maker may, for example, give a constraint that the population size should be at least the minimum viable population size. However, it is possible that the population size in the chosen solution is not viable due to the uncertainty concerning the value of the constraint. This can be prevented, if the uncertainty is taken into account using chance-constrained optimisation, by formulating the constraint $\sum_{j=1}^{n} a_{ij} x_j \geq b_i$ as (Hof & Pickens, 1991)

$$P(\sum_{j=1}^{n} a_{ij} x_j \geq b_i) \geq \varphi_i \, . \tag{7}$$

In this formula the probability of the solution to be feasible is set to φ_i, which may be, for instance, 95 %. This problem can be solved with the usual LP method, by taking a new constraint level $\tilde{b}_i$ from the 95 % percentage point of the cumulative distribution of b_i. Thus, the value of the constraint is increased in order to choose a solution that is on the safe side despite the uncertainty. If a critical population size is used as a constraint, the actual population size needs to increase with increasing uncertainty about the critical value (Haight & Travis, 1997).

Uncertainty about the parameters a, i.e. in how the population reacts to the treatment, can also be considered using chance-constrained optimisation. If the coefficients a are assumed to be normally distributed, constraint $\sum_{j=1}^{n} a_{ij} x_j \geq b_i$ can be presented as (Hof et al., 1992)

$$\sum_{j=1}^{n} \alpha_{ij} x_j + \delta_i \sqrt{\sum_{j=1}^{n} \sum_{h=1}^{n} x_j x_h \sigma_{ijh}^2} \geq b_i \tag{8}$$

where α_{ij} is the mean of a_{ij} and σ_{ijh}^2 is the covariance of a_{ij} and a_{ih}, and δ_i is the standard normal deviate corresponding to the required probability . This formulation follows from $\sum_{j=1}^{n} a_{ij} x_j$ being normally distributed with mean $\sum_{j=1}^{n} \alpha_{ij} x_j$ and variance $\sum_{h=1}^{n} \sum_{j=1}^{n} x_j x_h \sigma_{ijh}^2$. If the constraints are assumed to be mutually independent, any number of this type of constraints can be included in the problem formulation. The problem is a non-linear optimisation problem, which can be solved, for example, with simulation (Pickens et al., 1991) or with a so-called cutting plane-algorithm (Weintraub & Abramovich, 1995).

It can be required that not only each constraint is met with a given probability, but that a set of constraints is met with a given probability (joint probability chance constraint). Also, it is possible to maximise the probability for meeting the constraints instead of using a constant probability (Hof & Pickens, 1991, Hof et al., 1992).

The above mentioned formulation is suitable, if the uncertainty involved is random variation. If the uncertainty is due to ignorance or ambiguity, this formulation does not apply. If the value of critical threshold, for example, is an expert opinion, the threshold may be given as an interval value. Then, fuzzy linear programming may be utilised (e.g. Mendoza & Sprouse, 1989, Bare & Mendoza, 1992, Mendoza et al., 1993). The fuzzy optimisation is presented in more detail in the next chapter, in the context of optimisation based on the use of utility function.

3.2. Optimisation based on utility function

If, instead of the LP method, a utility function approach is used in planning, each variable, including the size of the population or the viability of the population, has to be taken into account in the utility function. HERO heuristic optimisation method (Pukkala & Kangas, 1993) is an example of approaches based on multi-attribute utility theory. It has been developed especially for tactical forest planning at area level. With HERO, optimisation may be divided into two main phases: (1) estimation of the utility model, and (2) maximisation of the utility model. Although HERO was originally developed for planning of private non-industrial forestry with a single decision-maker, the method can be applied as well with multiple decision-makers and with public participation (Kangas et al., 1996).

The initial version of HERO was constructed on the premise of an additive utility model:

$$U = \sum_{i=1}^{p} a_i u_i(q_i), \tag{9}$$

where U is the total utility, p is the number of objectives, a_i is the relative importance of objective i, $u_i(q_i)$ is the partial utility function of objective i, and q_i is the amount that the plan produces or consumes objective variable i.

When estimating the weights, the relative importance of the objectives is compared in a pairwise manner. The weights can be estimated based on the comparisons using, e.g., the eigenvalue technique (Saaty, 1977) or regression-based methods as described by Alho and Kangas (1997). The sub-priority functions u for each objective variable i, are estimated with the aid of the predicted production possibilities, using techniques similar to the procedure in estimating the relative importances. There are no preconditions on the form of a sub-priority function. In the maximisation, a heuristic direct-search algorithm is used to search for the best combination of standwise treatment schedules for the forest area.

If the objectives of the decision-maker are not complementary, (i.e. good gain in one objective cannot compensate for an inferior result in another) an additive utility function may not be suitable. For example, if the population size is considered as a part of an additive utility function, there is no guarantee that it is above zero in the optimal treatment schedule. In such a case, a multiplicative model may be more appropriate. Furthermore, by means of a multiplicative part, strict restrictions can be dealt with (like constraints in mathematical programming). Multiplicative parts can be included in the HERO utility model since a heuristic algorithm is used in the maximisation phase.

If the sizes of r wildlife populations are taken into account as multiplicative parts U_j of an utility function, as

$$U = \left(\sum_{i=1}^{p} a_i u(q_i) \right) \prod_{j=1}^{r} U_j \tag{10}$$

the size of each population has to be above zero in the optimal solution. (Otherwise the utility of each management alternative is zero.) If the population viability can be expressed as a continuous function of forest characteristics, this function can also be used as a partial utility function.

A crisp constraint can be met by defining U_j as a threshold function R (Kangas & Kangas, 1998)

$$R = \begin{cases} 0 & if \;\; S_i < S_{crit} \\ 1 & otherwise \end{cases}, \tag{11}$$

where S_i is the actual value of critical variable S in alternative i and S_{crit} is the threshold value (Fig. 1). When threshold function R is used in a multiplicative part of a utility function, all the treatment schedules not meeting the given requirements are rejected. Thus, a minimum population size can be guaranteed in an optimal solution by using a threshold function of the population size as a multiplicative part of a utility function. This threshold function takes value zero if the population size is below a given threshold and value one if the size is at least the given threshold. Similarly, the threshold function could be formed so that the value of the function is zero if the population is not viable (for example, with 95% probability), and one otherwise.

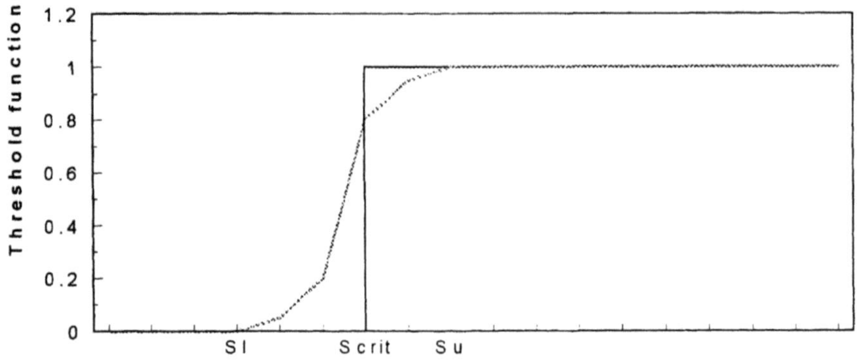

Figure 1. The threshold function value with respect to critical variable S.

Of course, in this approach too the uncertainty needs to be accounted for. For example, different models for the population development or different experts could give different threshold values with respect to the population viability. Even if statistical uncertainty assessment for the threshold value is not available, it is often possible to assume the true threshold to range within a given interval. Using the upper bound of the interval as a value of the threshold in the utility function means risk-aversing policy under uncertainty, and corresponds to the precautionary principle. Consequently, using the lower bound of the interval means risk-seeking policy. Uncertain threshold values for the population size or viability can also be utilised by forming a fuzzy threshold function. The fuzzy threshold function can be defined as (e.g. Mendoza et al., 1993)

$$R = \begin{cases} 0 \ \ if\, S_i < S_l \\ \dfrac{S_i - S_l}{S_u - S_l} \ \ if\, S_l \le S_i \le S_u \\ 1 \ \ if\, S_i > S_u \end{cases} \tag{12}$$

where S_u is the upper bound of the threshold value and S_l is the lower bound of it. Thus, the optimal solution will be found among the solutions which fulfil the minimum threshold value. The fuzzy threshold function may also be defined non-linear (Fig 1). If (subjective) probabilities of the possible threshold values can be assessed, the function R may reflect these probabilities.

It is not enough to consider the risks in a single point in time. The management alternative which minimises the risks in the short run, may not be the best in the long run. The whole time profile of survival probability can be taken into account in a utility function presented by McKelvey (1996)

$$U(T) = \sum_{t=0}^{T-1} \pi(t) \tag{13}$$

where T is the extinction time and $\pi(t)$ ($\pi(t) > 0 \ \forall \ t < T$) is defined as a momentary value of survival probability during a time-step t. The value of this utility function is the larger, the more the extinction can be delayed. It is possible to weigh the risks in the near future more than the risks in the distant future (McKelvey, 1996).

In some cases it may be sufficient to assume the environment as static, so that the effect of management alternatives is permanent. Such may be the case, when an old-growth forest is cut and transformed to an urban area. Usually however, the forest area changes dynamically all the time and the changes are reversible within a certain time period. Since the time horizon is long, it is often possible to adapt to the situation at hand, i.e. to change the decisions when the future unfolds and new information becomes available (e.g. Gong 1998). Although it is not possible to replace a clear-cut old-growth forest, it is possible to reduce the clear-cut area somewhere else. On the other hand, the effect of treatments on the wildlife populations may be lagging. Thus, dynamic conditions in the whole time horizon need to be accounted for. Optimally, a predictive distribution of the population would be available for each year and each treatment alternativey. Unfortunately, such an information is usually not available.

Consequently, the whole time horizon needs to be considered when using the threshold function approach. For example, if the critical levels of forest conditions are considered only at the end of the planning period, the species of interest may have become extinct in the interim. In a threshold function defined above, the threshold was assumed to be constant irrespective of the time-horizon considered. It is, however, probable that a threshold value for surviving the next 10 years would be less restrictive than that for the next 100 years. Thus, a threshold function should be defined such that it prefers a reversible management alternative (temporary decline in habitat quality) over an alternative whose effects are irreversible from the population's point of view.

The dynamic nature of the problem can be considered by constructing a separate threshold value for each sub-period. Such sub-periodical threshold values could be used in many ways, reflecting different policies with respect to risk. An ad-hoc solution, assuming a threshold function for a predefined time horizon available, is to define the total threshold function value as a product of sub-period values R_i.

$$R = \prod_{i=1}^{t} R_i \tag{14}$$

and assuming constant habitat quality through the whole time horizon, the sub-period values are $R_i = \sqrt[t]{R}$.

If the threshold function is assumed crisp, this approach gives a threshold function value that corresponds to the smallest amount of suitable habitat in the planning period. This leads to a cautionary policy; the reversibility of the changes is not taken into account. If the threshold function is assumed fuzzy, some decline in the habitat area or quality is tolerated, the amount of tolerable decline depending on the shape of the threshold function. The threshold function, constructed as a product of sub-period values, gives higher priority for alternatives with temporary decline than to those with permanent decline.

4 An example

In the example it is assumed that the utility of the decision-maker consists of net present value of incomes (discount rate 5%, value of timber 200 FIM/m^3), the amount of timber at the end of the planning period and the viability of the Capercaillie population. Net incomes and amount of timber form the additive parts of the utility function, with weight 0.5 assigned to both objectives. The viability of Capercaillie is considered via a threshold function, which forms a multiplicative part of the utility function.

It is assumed that 10% of the stands above 60 years, 50 % of stands above 70 years and 95% of the stands above 80 years are suitable for Capercaillie. The critical proportion of suitable habitat is presumed to be 25% (S_{crit}, see Fig 1) in a fine-grained habitat (Rolstad & Wegge 1989). A threshold function R, defined as a function of the proportion of suitable habitat, is applied as a multiplicative partial-utility function (Eqs. 10 and 11). Two different threshold functions concerning the viability of Capercaillie are applied. In the first case, the capercaillie population is assumed to be viable if the proportion of suitable habitat is above the crisp critical value. In the second case, the threshold function is assumed to be fuzzy, with the true threshold value between 15%-35% (S_l and S_u, see Fig 1) , with 20% probability that it is larger than 25%.

The planning period is 50 years, divided to five 10-year sub-periods. The two forms of threshold function are calculated either at the end of the planning period, or the threshold function values are considered separately for each sub-period (eq. 14). The planning area consists of 300 hectares of forest area. The forest area is divided to 11 age classes: 0-10, 10-20, ... , 100+ years. The amount of timber in any area is assumed to depend only on the age of the trees in that area. The only treatment considered is clear-cutting the forests belonging to the oldest age-class. The initial characteristics of the area are presented in Table 1.

Age class	Area, ha	Volume, m^3/ha
0-10	40	0
10-20	10	39
20-30	10	74
30-40	50	108
40-50	20	141
50-60	10	174
60-70	10	206
70-80	10	238
80-90	30	269
90-100	40	300
100+	60	331

Table 1. The initial situation of the forest area in the example

The problem is to decide, how large an area is clear-cut in each 10-year period. A number of possible treatment schedules are considered, with different clear-cutting areas assigned to each sub-period. The treatment schedules considered are presented in Table 2, and the proportion of suitable habitat that results from applying each regime are presented in Table 3. The values of threshold functions in each treatment regime are presented in Table 4.

Regime Period	1	2	3	4	5
1	40	30	30	10	20
2	30	40	30	30	20
3	20	0	30	40	40
4	40	40	0	50	20
5	0	50	50	40	10
6	30	50	30	20	20
7	50	10	10	40	40
8	30	50	40	20	10

Table 2. The clear-cut area (ha) in each treatment regime and planning period

Regime Period	1	2	3	4	5
1	35	29	25	32	35
2	38	29	25	25	28
3	41	45	41	38	35
4	35	25	31	25	28
5	48	35	25	22	28
6	38	25	22	25	28
7	32	32	35	32	28
8	38	25	18	22	28

Table 3. The proportion of suitable habitat (%) in each treatment regime and planning period

	Crisp threshold		Fuzzy threshold	
Regime	End	Periods	End	Periods
1	1	0	0.998	0.927
2	1	0	0.898	0.876
3	1	1	0.998	0.999
4	1	1	0.898	0.893
5	1	0	0.898	0.793
6	1	0	0.898	0.744
7	1	1	0.898	0.959
8	1	0	0.898	0.547

Table 4. The value of different threshold functions in each treatment regime, considering the situation in the end of period five or in all periods.

In each of the four cases, a different treatment regime is chosen as optimal (Table 5). If only the situation at the end of the planning period is considered, the 25% proportion of suitable habitat is obtained with each regime. Then, using the crisp threshold function value, the regime with the largest net incomes and timber volume is chosen (regime 8). If the level of suitable habitat is considered separately in each sub-period, the regimes with the proportion of suitable habitat below 25% (4) during any of the planning periods are rejected. From the rest of the alternatives, the alternative in which net incomes and timber volume are largest is chosen. If the threshold function is the fuzzy one, and the situation is considered only at the end of the planning period, the two regimes with the largest proportion of suitable habitat also have the largest threshold function value. From these regimes, the one with larger net incomes and timber volume is chosen (1). If the proportion of suitable habitat is considered in each sub-period, the chosen regime is the one that maintains the proportion of suitable habitat near the 35% level through the whole planning period, but at the same time gives relatively good net incomes (7).

	Crisp threshold		Fuzzy threshold	
Regime	End	Periods	End	Periods
1	1968	0	**1965**	1824
2	1985	0	1783	1738
3	1223	1223	1221	1222
4	2071	**2071**	1860	1849
5	1584	0	1423	1256
6	2084	0	1872	1550
7	1963	1963	1764	**1883**
8	**2146**	0	1927	1171

Table 5. The value of total utility/1000. In each treatment regime with different threshold function values, considering the situation at the end of period five or in all periods.

5 Discussion

It has often been presumed that information about wildlife populations is too limited to be effectively used in forest management planning. Information about wildlife populations can, however, be used in forest management planning in several ways, and it need not necessarily be complete. Empirical models or species-wise information are not required and the planning may be based on expert knowledge as well. It is also possible to combine expert judgement and empirical information.

Models describing population dynamics can be used to deal with population sizes in optimisation. If the models are considered too complicated for planning purposes, or if the population size per se is not considered a relevant decision variable, the information can be utilised in risk analyses. The results of risk analysis, for example in the form of critical thresholds of certain forest characteristics, can then be used in searching for means to maintain or increase the population viability.

The spatial aspects in forest management planning have not been considered in this paper. However, including objectives which concern habitat connectivity or habitat patch size distribution introduces new requirements for the planning models. The planning problems including spatial considerations may be very complicated, and heuristic methods are often needed (see e.g. Sessions 1992; Öhman & Eriksson 1998).

Information about wildlife populations is often presumed too uncertain to be useful in planning. However, uncertainty should not preclude planning: it is not wise to ignore available information because of underlying uncertainty. Plans based on no information at all are not much better than plans based on uncertain information, provided the uncertainty is not ignored. Nevertheless, the uncertainty has to be taken into account in planning, in order to obtain useful plans. Uncertainty in ecological considerations has two main sources: variability, which is due to the stochastic nature of the system, and ignorance, which is due to incomplete knowledge about the system (e.g. Ferson & Ginzburg, 1996). Both of these sources need to be accounted for in planning, and in the most in-depth analysis their effects should be distinguished.

Uncertainty involved in ecological (or other) considerations can be taken into account by assessing the probabilities of possible outcomes, and then comparing alternative plans according to this information. Optimally, the probability of any one plan being better than the others, with respect to the objectives set, can be calculated (Alho & Kangas, 1997). Different decision-makers have different attitudes towards risk and uncertainty. The optimal plan for a risk taker often differs from that of a risk avoider. This is also the case when considering ecological assessments. In modern decision support, the decision-makers' risk preferences can also be taken into account (e.g. Pukkala & Kangas, 1996). †

References

Alho, J., Kangas, J. & Kolehmainen, O., 1996: Uncertainty in expert predictions of the ecological consequences of forest plans. *Applied Statistics* 45: 1-14.

Alho, J., & Kangas, J., 1997: Analyzing uncertainties in experts' opinions of forest plan performance. *Forest Science* 43: 521-528.

Bare, B.B. & Mendoza, G.A., 1992: Timber harvest scheduling in a fuzzy environment. *Canadian Journal of Forest Research* 22: 423-428.

Burgman, M.A., Church, R., Ferguson, I., Gijsbers, R., Lau, A., Lindenmayer, D., Loyn, R., McCarthy, M. & Vanderberg, W., 1994: Wildlife planning using FORPLAN: a review and examples from Victorian forests. *Australian Forestry* 57: 1131-1140.

Burgman, M.A., Ferson, S. & Akçakaya, H.R., 1993: *Risk assessment in conservation biology*. Population and Community Biology Series 12. Chapman & Hall. 314 p.

Draper, D., 1995: Assessment and propagation of model uncertainty. *Journal of Royal Statistical Society, Series B* 57: 45-97.

Drechsler, M., Burgman, M.A. & Menkhorst, P.W., 1998: Uncertainty in the population dynamics and its consequences for the management of the orange-bellied parrot *Neophema Chrysogaster*. *Biological Conservation* 84: 269-281.

Caughley, G., 1994: Directions in conservation biology. Review. *Journal of Animal Ecology* 63: 215-244.

Ginzburg, L.R., Ferson, S & Akçakaya, H.R., 1990: Reconstructibility of density dependence and the conservative assessment of extinction risks. *Conservation Biology* 4: 63-70.

Gong, P., 1998. Risk preferences and adaptive harvest policies for even-aged stand management. *Forest Science* 44:496-506.

Faucheux, S. & Froger, G., 1995: Decision-making under environmental uncertainty. *Ecological Economics* 15: 29-42.

Ferson, S. & Ginzburg, L.R., 1996: Different methods are needed to propagate ignorance and variability. *Reliability Engineering and System Safety* 54: 133-144.

Haight, R.G. & Travis, L.E., 1997: Wildlife conservation planning using stochastic optimisation and importance sampling. *Forest Science* 43: 129-139.

Hanski, I., 1989: Metapopulation dynamics: does it help to have more of the same? *Trends in Ecology & Evolution* 4: 113-114.

Hanski, I., 1991: Single-species metapopulation dynamics: concepts, models and observations. *Biological Journal of the Linnean Society* 42: 17-38.

Hanski, I. & Gilpin, I., 1991: Metapopulation dynamics: brief history and conceptual domain. *Biological Journal of the Linnean Society* 42: 3-16.

Harrison, S. & Quinn, J.F., 1989: Correlated environments and persistence of metapopulations. *Oikos* 56: 293-298.

Hof, J.G. & Pickens, J.B., 1991: Chance-constrained and chance-maximizing mathematical programs in renewable resource management. *Forest Science* 37: 308-325.

Hof, J.G., Kent, B.M. & Pickens, J.B., 1992: Chance constrained and chance maximisation with random yield coefficients in renewable resource optimisation. *Forest Science* 38: 305-323.

Kangas, A. & Kurki, S., 2000: Predicting the future of Capercaillie (*Tetrao Urogallus)* in Finland. Submitted manuscript.

Kangas, J., 1992: *Metsikön uudistamisketjun valinta - monitavoitteiseen hyötyteoriaan perustuva päätösanalyysimalli*. Summary: Choosing the regeneration chain in a forest stand: A decision-model based on multi-attribute utility theory. Joensuun yliopiston luonnontieteellisiä julkaisuja 24. 230 s.

Kangas, J., 1994: Incorporating risk attitude into comparison of reforestation alternatives. *Scandinavian Journal of Forest Research* 9: 297-304.

---

† The authors' address: Finnish Forest Research Institute, Kannus Research Station, P.O. Box 44, FIN-69101 Kannus, Finland. The date of preparation: 29th May 2000.

Kangas, J., Karsikko, J., Laasonen, L. & Pukkala, T., 1993: A method for estimating the suitability function of wildlife habitat for forest planning on the basis of expertise. *Silva Fennica* 27: 259-268.

Kangas, J., Loikkanen, T., Pukkala, T. & Pykäläinen, J., 1996: A participatory approach to tactical forest planning. *Acta Forestalia Fennica* 251. 24 p.

Kokko, H., Lindström, J. & Ranta, E., 1997: Risk analysis of hunting of seal population in the Baltic. *Conservation Biology* 11: 917-927.

Kokko, H. & Lindström, J., 1998: Seasonal density dependence, timing of mortality, and sustainable harvesting. *Ecological Modelling* 110: 293-304.

Lamberson, R., McKelvey, R., Noon, B. & Voss, C., 1992: A dynamic analysis of spotted owl viability in a fragmented forest. *Conservation Biology* 6: 505-512.

Lande, R., 1993: Risks of population extinction from demographic and environmental stochasticity and random catastrophes. *The American Naturalist* 142: 911-927.

Leskinen, P. & Kangas, J., 1998: Analysing uncertainties of interval judgement data in multiple-criteria evaluation of forest plans. *Silva Fennica* 32: 363-372.

Leslie, P.H., 1945: On the use of matrices in certain population mathematics. *Biometrika* 33: 183-212.

Marshall, K. & Edward-Jones G., 1998: Reintroducing capercaillie (*Tetrao urogallus*) into southern Scotland: identification of minimum viable populations at potential release sites. *Biodiversity and Conservation* 7: 275-296.

McCarthy, M.A., Pearce, J.L. & Burgman, M.A., 1994: Use and abuse of wildlife models for determining habitat requirements of forest fauna. *Australian Forestry* 57: 82-85

McCarthy, M.A., Burgman, M.A. & Ferson, S., 1995: Sensitivity analysis for models of population viability. *Biological Conservation* 73: 93-100.

McCarthy, M.A., Burgman, M.A. & Ferson, S., 1996: Logistic sensitivity analysis and bounds for extinction risks. *Ecological Modelling* 86: 297-303.

McCarthy, M.A., 1996: Extinction dynamics of the helmeted honeyeater: effects of demography, stochasticity, inbreeding and spatial structure. *Ecological Modelling* 85: 151-163.

McKelvey, R., 1996: Viability analysis of endangered species: a decision theoretic perspective. *Ecological Modelling* 92: 193-207.

Mendoza, G.A., Bare, B.B. & Zhou, Z., 1993: A fuzzy multiple objective linear programming approach to forest planning under uncertainty. *Agricultural Systems* 41: 257-274.

Mendoza, G.A. & Sprouse, W., 1989: Forest planning and decision-making under fuzzy environments: an overview and illustration. *Forest Science* 35: 481-502.

Mills, L.S., Hayes, S.G., Baldwin, C., Wisdom, M.J., Citta, J., Mattson, D.J. & Murphy, K., 1996: Factors leading to different viability predictions for a grizzly bear data set. *Conservation Biology* 10: 863-873.

von Neumann, J. & Morgenstern, O., 1947: *Theory of games and economic bahavior*. John Wiley. New York. 641 p.

Öhman, K. & Eriksson, L.O., 1998: The core area concept in forming contiguous areas for long-term forest planning. *Canadian Journal of Forest Research* 28:1032-1039.

Pickens, J.B., Hof, J.G. & Kent, B.M., 1991: Use of chance-constrained programming to account for stochastic variation in the A-matrix of large-scale linear programs. A Forestry application. *Annals of Operations Research* 31: 511-526.

Pukkala, T. & Kangas, J., 1993: A heuristic optimisation method for forest planning and decision-making. *Scandinavian Journal of Forest Research* 8: 560-570.

Pukkala, T. & Kangas, J., 1996: A method for integrating risk and attitude towards risk into forest planning. *Forest Science* 42: 198-205.

Raftery, A.E., Givens, G.H. & Zeh, J.E., 1995: Inference from a deterministic population dynamic model for bowhead whale. *Journal of American Statistical Association* 90: 402-415.

Reed, W.J., 1993: The decision to conserve or harvest old-growth forest. *Ecological Economics* 8: 45-69.

Reed, M.J., Elphich, C.S. & Oring, L.W., 1998: Life-history and viability analysis of the endangered Hawaiian stilt. *Biological conservation* 84: 35-35.

Rolstad, J, & Wegge, P., 1989: Capercaillie *Tetrao urogallus* populations and modern forestry - a case for landscape ecological studies. *Finnish Game Research* 46: 43-52.

Saaty, T.L., 1977: A scaling method for priorities in hierarchical structures. *Journal of Mathematical Psychology* 15: 234-281.

Sæther, B-E., Engen, S., Islam, A., McCleery, R. & Perrins, C., 1998: Environmental stochasticity and extinction risk in a population of a small songbird, the great tit. *The American Naturalist* 151: 441-450.

Salwasser, H., Mealey, S.P. & Johnson, K., 1984: Wildlife population viability: a question of risk. *Trans. N. Amer. Wildl. and Natur. Resour. Conf.* 49: 421-439.

Schaffer, M.L., 1981: Minimum population sizes for species conservation. *Bioscience* 31: 131-134.
Sessions, J., 1992. Solving for habitat connections as a Steiner network problem. *Forest Science* 38:203-207.
Swart, J. & Lawes, M.J., 1996. The effect of habitat patch connectivity on samango monkey (*Cercopithecus mitis*) metapopulation percistence. *Ecological Modelling* 93:57-74.
Taylor, R.J. & Haseler, M., 1993. Occurrence of potential nest trees and their use by birds in sclerophyll forest in north-east Tasmania. *Australian Forestry* 56:165-171.
Weintraub, A. & Abramovich, A., 1995: Analysis of uncertainty of future timber yields in forest management. *Forest Science* 41: 217-234.

RISK APPRAISAL FOR FOREST MANAGEMENT WITH RESPECT TO SITE QUALITY AND ENVIRONMENTAL CHANGES

Michael Bredemeier[1], *Norbert Lamersdorf*[2], *Hubert Schulte-Bisping*[2] *and*

Burckhardt von Lüpke[3]

ABSTRACT

Environmental changes and human interference have substantially modified forest site conditions in Europe for millenia. We illustrate how site conditions and ecosystem structure in forests were affected and frequently deteriorated to the point of severe degradation by human settlement and exploitation. Environmental changes, human interferences and the subsequent ecosystem transformations are varied and operating on different spatial and temporal scales, depending on the processes and responses under consideration. A particular characteristic of recent anthropogenic environmental changes is that they presumably occur much more rapidly than natural variations, and that they may lead to constellations of ecological effects that are unprecedented in history and evolution. Unusual combinations of growth and vitality factors arise from recent environmental changes, some factors under change bearing the potential to foster vitality and growth, e.g., CO_2 *enrichment and enhanced nitrogen deposition, while others may have strong detrimental effects, like soil acidification, compaction or loss by erosion. We quote a nationwide investigation of forest soil status in Germany, revealing that the majority of forest soils are indeed affected by acidification, at least in the upper soil layers which are most important for rooting. Appropriate silvicultural strategies are capable of significantly reducing risk with respect to many of the contributing factors. More diverse forest ecosystems mimicking structures and functions in natural forests are expected to bring about more stability. The classical risk formula calculates risk as the numerical product of probability and amount of damage. For such a quantification in forestry, however, probability of damage is altogether uncertain, at least at the individual stand or site level. Nevertheless, qualitative assessments can be performed with some confidence, and they should be performed before making management decisions, which will influence forest structure and functions for decades to centuries.*

Keywords: *risk assessment, forest management, silviculture, soil acidification, site quality, soil degradation, scales*

---

[1] Forest Ecosystems Research Centre, Univ. of Goettingen, F.R.G., email: mbredem@gwdg.de

2 Institute of Soil Science & Forest Nutrition, Univ. of Goettingen, F.R.G., email: nlamers@gwdg.de

3 Institute of Silviculture, Univ. of Goettingen, F.R.G., email: bluepke@gwdg.de

K. von Gadow (ed.), Risk Analysis in Forest Management, 21–48.

1 Introduction

Forestry is characterized by long perspectives, much longer than for most other land-use systems. Even in plantation forestry with fast growing species, or agroforestry, rotations of trees are at least several years to a few decades long, while agricultural systems have rotations of one to multiple crops per year. "Classical" forestry schemes developed by forest science in Europe over the past two centuries are aiming at high quality timber production in large bole dimensions and therefore employ time spans of typically one to two centuries, depending on growth characteristics of the species selected and on site conditions. The latter have in the past always been considered constant over arbitrarily long time horizons. The reliance on this constancy is expressed in the yield class tables used by foresters, as compiled for Germany by Schober (Schober, 1975). Yield class tables are empirical models based on a large number of test sites, describing the average relationship of growth and site productivity, categorized in ranked yield classes.

The validity of established yield class tables has recently become dubious in the light of substantial deviations of actual tree growth in Europe from such models (Spiecker, 1999; Beese, 1996). Observations indicate that forest growth was enhanced in many locations, corresponding to a positive shift in site class of many stands. On the other hand, growth may decline where acute forest damage occurs. All these effects are probably caused by changes in site conditions over roughly the past century due to anthropogenic influences, particularly atmospheric deposition. Atmospheric deposition has two prominent aspects, the acidification aspect and the nutrient input aspect. While acidification of forest soils triggers several mechanisms that impair forest nutrition and ecosystem resilience (Ulrich, 1983; Ulrich, 1994), deposition of nutrients, particularly nitrogen, may foster tree growth (Spiecker, 1999; Binkley and Hogberg, 1997). However, ecological risks arising from nitrogen eutrophication of forest ecosystems are at odds with potential gains in tree growth. Transfer of excess nitrogen to aquatic and groundwater systems in the form of nitrate is generally considered the most serious risk arising from "nitrogen saturation" (Aber et al., 1989). Additional risks identified comprise unbalanced tree nutrition and the transfer of nitrogenous greenhouse gases to the atmosphere (Brumme et al., 1999; Rennenberg et al., 1998; Papen and Butterbach-Bahl, 1999).

Other environmental changes than acidification and eutrophication will also play a role in determining future forest growth conditions, particularly the ongoing increase in atmospheric carbon dioxide concentrations, a possible climatic change with a higher frequency of extreme weather conditions, and changes in silvicultural and management practices in forests.

Consequently, forest managers can no longer rely on site quality as a constant factor in their long-term planning (it is an open question if they ever should have). They have to consider ongoing and possible future environmental changes and assess risks and chances connected to their management decisions over long time scales.

In the present paper we briefly refer to a simple and generally accepted concept of risk calculation, and discuss the limits of its application in forest management. We then give an overview of environmental changes relevant for forest ecosystems and forestry, structured by the spatial and temporal scales on which they operate. Following that we illustrate the actual soil chemical status of forest sites in Germany based on an intense large-scale inventory and an example case study, and discuss its implications for risk appraisement. Finally, we report new silvicultural concepts and strategies for improving site conditions and reduce risk in German state forest management.

2 The Risk Formula

A widely known symbolic representation to calculate risk is given by the formula

$$R = p_d * A_d,$$

where R is risk, p_d is the probability of damage to occur, and A_d is the amount of damage, usually expressed as a monetary equivalent (Braess, 1960; Banse, 1996).

While the above algebraic expression and its underlying concept are simple and concise in many instances, it can not be resolved quantitatively. The formula originales from the field of insurance, where precise quantitative information on damage amounts and probabilities of damage are available (the latter e.g. as the proportion of accidents in the total number of insured cars).

However, quantitative information of such kind is unavailable, when risk has to be assessed for a single case, e.g., a single forest site, rather than for a large, statistically well-evaluated collective. Risk at the forestry management unit level can only be estimated with a considerable degree of uncertainty. A_d, the potential amount of damage, can be estimated with some confidence, at least with respect to losses of timber value. However, p_d, the probability of damage, is highly uncertain, since damage events in forestry are most often related directly (storms, drought, frost) or indirectly (insects) to weather conditions and/or other fluctuations (e.g., pathogen population dynamics). Monetary damage equivalents are hard to assess when damage to site quality or ecological resources are concerned, e.g., soil loss by erosion or pollution of water following forest disturbance.

Thus, assessment of p_d in forest planning and management will remain subject to uncertainty, with the consequence that they have to be of a more qualitative rather than quantitative nature. Nevertheless, awareness of the relevant factors contributing to damage probability is absolutely essential for the planning process in forest management, even if such awareness is merely qualitative. In the following sections, we give an overview of those environmental changes that affect forest site conditions, and active forest management strategies to improve forest ecosystem stability. Economic and engineering aspects of the risk problem are covered in other papers of this volume.

3 Environmental changes pertinent to multiple scales

Environmental changes relevant to forest management are reflected in changing site conditions over time. The onset, intensity and duration of such changes, as well as their spatial extent can be very different for the various processes of environmental change. In this section we use the differentiation of scales to structure our analysis of those environmental changes, which have a prominent effect on risk in forest management.

3.1 Local to regional scales: historical over-exploitation and degradation

Under background conditions without human influence, forests are the dominant land cover in the tropical, temperate and large parts of the boreal climate zones. In Germany, for instance, the "potential natural vegetation" would be essentially all forest except for bogs and swamps in the lowlands and sites above the timberline in the high mountains (Ellenberg, 1988). However, in the regions of the globe settled by humans, natural primary forests have mostly been cleared and replaced by managed production forests and degradational states. The characteristics of managed production forests range from quasi-natural secondary growth to artificial plantation schemes with imported species and high fertilizer and energy inputs.

With the exploitation and management of forests in historical times, humans have also changed the biogeochemical boundary conditions of forest ecosystems under their control. In many regions of Europe overexploitation of forests, often leading to devastation, was a common feature. Site quality (e.g., nutrient stores and buffer capacity) substantially deteriorated, probably even stronger than by the more recent processes of atmospheric deposition discussed in the next section.

It is important to note that not only the excessive harvest of timber led to overexploitation in past centuries, but also other forms of forest utilization, which exerted additional – and sometimes even stronger – detrimental effects on site quality. The major mechanisms by which additional exploitations deteriorated forest and soil were:

- additional export of nutrients, often exceeding the export caused by timber utilization. Besides resulting in strongly reduced nutrient stores, this may result in an intense soil acidification effect, which considerably decreases the further resilience of the forest ecosystem (Ulrich, 1981).
- obstruction of natural regeneration by domestic animal feeding (e.g., cattle feeding on saplings and young trees as well as pigs feeding on acorns)
- deterioration (compaction and erosion) of soil by cattle feeding in the forest

Most prominent among the practices referring to the first point above was litter raking, i.e., the export of leaf litter and surface humus for use in farm stables. The forest litter was brought into the stables as a substitute for straw, there it was mixed and enriched with animal manure, and afterwards was ploughed into the fields. Hence, the whole process was an export of nutrients and humus (the latter being important for soil structure) from the forests to the arable land. This led to an improvement of fertility in arable soils, that could only be achieved in this way, since modern fertilizers were not yet available. Even the access to natural materials like limestone for soil improvement was not possible in areas where such resources were too far away. The moderate gains in fertility of field soils, however, were paid for by considerable impoverishment of forest sites.

Almost throughout Europe, the forests were also used as a pasture for cattle, goats and pigs. This practice led to disturbances of natural regeneration. Moreover, it is likely that erosion by water occured on steeper slopes, where cattle disturbed the surface soil. Twigs and leaves were also taken from living and harvested trees and used as fodder in times when other fodder sources were scarce. This form of exploitation not only increased nutrient exports from the forests, but it contributed to a further deterioration of the forest structure.

The historical (over)exploitation of forests gravely impaired site quality. Typical final states of the forest after long-term use of these practices were open, degraded forest landscapes with only a few old trees remaining, or expanded heathlands with a dwarf shrub cover on highly impoverished soil. Many old paintings from the Middle Ages up to the 18th century show such states of the landscape very clearly. Fig. 1 is a very old example which dates from the beginning of the 14th century, depicting the clearing of trees and their utilization for construction and support of human settlements (the feudal rule of such developments can also clearly be recognized, it is in fact the main subject of the drawing).

Figure 1. A drawing dating from the beginning of the 14th century showing the felling of trees and their utilization for construction of settlements under a nobleman's rule. From Plochmann 1979, original in the National Museum of Bavaria, Munich.

While site quality and forest ecosystem structure suffered impairment during centuries of exploitation, the modern European land-use pattern and in fact modern civilization were formed during those days. Forest management in general was not sustainable. Rural and urban ecosystems were developed and sustained at the expense of forest ecosystems. While the overall time scale of non-sustainable forest exploitation in Europe extends over more than a millenium, spatial scales of interference are smaller and of a more individual nature, depending on local or regional traditions and preferences in land-use. Therefore, it is eminently important to consider site history in appraising actual site quality and potential in forest planning and risk assessment at the forest management unit level.

3.2 Regional to continental scales: shifting of forest site conditions caused by long-range air pollution

In a more recent chapter of history, the industrialization period, new stresses to forest ecosystems and their soils occured. Air pollution close to emission sources occasionally reached such levels that direct damage to forest vegetation occured. This phenomenon was described as "smoke damages" (*Rauchschäden* in German) and has been reported early in the literature (Schroeder and Reuss, 1883; Wieler, 1905). Apart from such strong direct effects under high air pollutant concentrations, another process, slower but spatially more widespread, developed. Components of air pollution, chemically transformed into strong mineral acids, advanced soil acidification further than had already been attained by exploitation practices. Opposed to the soil acidification process by excessive organic matter export, acid deposition introduced large amounts of mobile, conservative anions (predominantly sulfate) into the soil water, thus facilitating aquifer- and surface water acidification. This feature of anthropogenic acid deposition is equally or probably even more important than the further contribution of acid rain to forest soil solid phase acidification. Comprehensive and detailed reviews of the chemical processes involved in soil and water acidification can be found (e.g. in Möhring, 1992; Schweiger, 1989; Ulrich, 1972).

It is noteworthy that the period of industrialization and augmentation of acid deposition coincided with the development of scientifically based, modern forestry, which put an end to overexploitation and was successful in recuperating and sustaining yields of timber from forests. With the end of immoderate utilization practices, a possibility of recovery was given for forest soils, at least to some extent. So there were presumably two opposing trends operating at the same time: recovery of forest structure and of soil organic matter on the one hand, offering some potential for a slow recovery of soil structure and of nutrient stores. On the other hand, acid deposition played an important role in deteriorating site quality, as was directly evidenced in some comparisons of chemical parameters in forest soils over long time periods (Falkengren-Grerup and Tyler, 1992);(Emmer et al. 2000). The extent of forest soil acidification in Germany will be demonstrated in section 4 below. It is generally assumed that acidification led to regional- to continental-scale destabilization of forest ecosystems in intensively industrialized regions (e.g., Central Europe, the Eastern U.S., Japan). A "novel type forest decline" (German: *neuartige Waldschäden*), caused by manifold and complex stress mechanisms related to acidification was discovered (Schmieden and Wild, 1995; Ortloff and Schlaepfer, 1996).

A change in rooting patterns as a consequence of an altered soil chemistry has been demonstrated in numerous case studies (Godbold, 1994; Matzner and Murach, 1995). Long-term monitoring in some case studies, e.g. the German Solling Project, revealed a change in the trend in sulfur and acidity depositon beginning at the end of the 1980s (Meesenburg et al., 1995). The reasons for this change in the trend were air pollution legislation measures in West Germany and the closing of highly polluting industries in the eastern part of Germany and former Czechoslovakia (Umweltbundesamt, 1997). The mitigation of atmospheric acidity loads to forest soils offers an opportunity to stabilize element cycles in forest ecosystems. The potential for a relatively fast recovery of soil water chemistry was demonstrated in an ecosystem manipulation experiment with application of artificially prepared, "pre-industrial" precipitation (Bredemeier et al., 1998). However, the impaired rooting patterns and the strongly acidified soil solid phases (often with large amounts of acidity stored in unstable or meta-stable binding forms), represent an ecological long-term memory of acidification that can only be cancelled gradually (Godbold, 1994; BMELF, 1997).

It appears that root systems adapt to worsening soil chemical conditions by a loss of fine root biomass, a dieback of the root system in the deeper mineral soil, and a shallow rooting pattern, restricted to upper soil layers that are influenced by humic compounds (de-toxifying monomeric aluminium ions). Together with a distinct aluminium enrichment in deeper roots, these findings can be interpreted as evidence for aluminium toxicity. However, direct proof of Al-toxicity under field conditions is still lacking (Godbold, 1998; Godbold and Jentschke, 1998).

Another issue of concern besides soil acidity is the high atmospheric nitrogen deposition to forest ecosystems. In contrast to sulfur and acidity inputs, no decreasing trends can be discerned for N compounds yet (Persson and Majdi, 1995; Whytemare et al., 1997). Nitrogen inputs from the atmosphere can have a fertilizing effect and foster tree growth (Binkley and Hogberg, 1997), but they may also lead to nutritional imbalances giving rise to stress effects (Pearson and Soares, 1995; Richter et al. 1995; Ortloff and Schlaepfer, 1996). A further, significant potential of ecological risk emanates when excess nitrogen is converted to nitrate in N-saturated forest ecosystems, menacing aquatic ecosystem and water resources, or when N is converted to greenhouse gases and emitted (Durka et al., 1994; Nystrom et al., 1995; Brumme, 1995; Ortloff and Schlaepfer, 1996; Borken and Brumme, 1997; Brumme et al. 1999).

These trends and relationships persist together with an unbroken trend of rising CO_2-concentrations in the atmosphere and the threat of climatic changes. They are forming a constellation of rapid and complex environmental changes to which the forests in industrialized regions were hitherto not exposed during their evolution, the consequences of which can not be predicted with certitude. The patterns and processes discussed above refer preferentially to industrialized regions in developed countries, or those strongly developing at present, e.g., India and China. They operate rather uniformly at spatial scales larger than those for the historical land-use practices considered in the last section (i.e., regions to parts of continents). Temporal scales range from decades to almost two centuries, depending on the duration of industrialization.

3.3 Global scale: deforestation and soil degradation

Loss of forests and degradation of soils represent pressing ecological problems affecting large areas of the globe. It takes various shapes, depending on the climatic, geological and topographical conditions in the respective regions. However, in the vast majority of cases it seems to be related to past and present human activities (Oldeman, 1992), particularly those connected with land use, particularly cultivation and exploitation of land.

This problem is not recent, as a famous quote attributed to the Greek philosopher Plato may illustrate: "Attica (Athens) was no longer cultivated by true herdsmen, who made husbandry their business, and were lovers of honor, and of a noble nature. As a result Attica had become deforested, the soils depleted, and there are remaining only the bones of the wasted body – all the richer and softer parts of the soil having fallen away."

	land area	*arable land incl. cropland*	*pasture & grasslands*	*forest & woodland*
		[million ha]		
Africa	2964	188	853	761
North and Central America	2178	271	362	855
South America	1753	103	495	846
Asia	2679	469	800	535
Europe	473	136	80	158
Former USSR	2190	229	344	823
Oceania	845	51	428	200
World	**13082**	**1447**	**3362**	**4178**

Table 1. Land area and land use (FAO - Food and Agriculture Organization of the United Nations, 1995)

Roughly 9 billion ha or 69% of the 13 billion ha of total land surface are vegetated. Natural and semi-natural ecosystems in the form of forests and grasslands under different intensities of human mangement still account for the largest proportion, while ca. 1.4 billion ha or 11% are currently estimated to be intensively managed arable land (table 1).

Soil degradation may be human induced or dueto a natural process. Numerous examples of anthropogenic degradational processes are given in (WBGU, 1994); a pattern of autochthonous, completely natural soil and forest degradation dynamic was reported by Fölster (Fölster, 1992; Bredemeier and Fölster, 1998). A characteristic that defines soil degradation is that it negatively affects the land in its functions within ecosystems, i.e. the storage and cycling of water, energy, and nutrients become impaired. Moreover, soil material is often lost from the site of its origin, transported, and deposited elsewhere.

It is estimated that almost 2 billion hectares of land have undergone degradation processes, the most important ones being erosion by water and wind (ca. 1.65 billion ha, table 2). For comparison, the global area of arable land, forests and woodlands is ca. 5.7 billion hectares (table 1). Regions most severely affected by soil degradation are frequently those with a high human population density, or with a strong human impact (e.g. highly mechanized large-scale farming in the "dust bowl" of the U.S. Midwest).

Besides erosion processes, chemical and physical alterations may lead to degradation. While the area affected by the latter is much smaller than that of water and wind erosion (table 2), they are almost exclusively caused by human interference.

The major causes of soil degradation comprise land clearing and deforestation, overgrazing, agricultural mismanagement, export of soil nutrients in harvested crop and wood biomass, conversion of soils for urban and traffic constructions, local contamination by disposal of (toxic) wastes or large-scale pollution by atmospheric deposition, and irrigation of arid land in an unsustainable manner, bringing about salinisation of the soil.

The loss of organic matter in soils due to inappropriate management practices can be considered as a kind of chemical degradation.

Types of degradation	**Light**	**Moderate**	**Strong/extreme**	**Total**
Water erosion	343	527	224	1094
Wind erosion	269	254	26	549
Chemical degradation	93	103	43	239
- loss of nutrients	52	63	20	135
- salinisation	35	20	21	76
- contamination	4	17	1	22
- acidification	2	3	1	6
Physical degradation	44	27	12	83
	749	**911**	**305**	**1965**

Table 2: Types of soil degradation and areas affected [million ha] (Oldeman, 1992)

A further ecological effect may be that species diversity is diminished and often lost as soils are degrading.

Forest planning and management on highly degraded soils is an exceptionally difficult task, involving particularly high levels of risk. This has led to the constitution of an entirly new field in forest science and -practice, namely restoration or reclamation forestry (Various Authors, 1993).

After this excursion into global ecological problems relevant for forestry we return to take a closer look at site conditions in Germany and their implications for management risks.

4. Actual state of forest soils in Germany and implications for risks in management

Motivated by the decline symptoms in the most common tree species and under various site conditions in the early 1980s in Germany, there was a need for a systematically implemented chemical soil survey covering almost 11 million ha of German forests. Although a great deal of data from single case studies about the complex of soil acidification, atmospheric deposition and forest dieback symptoms were already available at that time, there was no spatially explicit, data based knowledge about the connection between soil conditions and forest health. Soil morphological surveys (site mapping), which are habitually performed by the forest administration are not adequate for a quantitative assessment of soil chemical site properties.

After the establishment of appropriate special methodology and an organizational board in the mid 1980s, guidelines were developed to solve problems of harmonization and to prepare a nationwide soil survey. Methods were documented and binding for all partners (see BMELF, 1994: Instructions for the nation-wide soil survey *BZE, Bundesweite Bodenzustandserhebung im Wald*). All activities were co-ordinated by the *German Federal Ministry for Food, Agriculture and Forest (BMELF).*

It was important to design the soil survey in such a way that it could accommodate the already existing, nation-wide 4x4 km grid for the annual crown density survey and to the 16x16 km grid of the European *Level I* monitoring programme installed in 1984. Finally it was decided to set up an observation grid of 8x8 km. A preceding study determined that this grid would represent the total forest area with respect to the most important variables *"soil type"*, *"texture"*, *"elevation"* and *"potential natural vegetation"* (Kuhnt et al., 1991, BMELF, 1997). Between 1987 and 1993, and thus including also the five new federal states of eastern Germany (former *GDR*), the following field information and analytical parameters were recorded for each of the 1800 measurement points:

- Standard soil survey parameters (i.e. bedrock material, soil type, humus form, elevation, tree species, specific treatments (e.g., liming), site history)
- Crown conditions of sample plots (about 30 trees per plot)
- Needle and leaf analysis for major nutrients and some trace elements (Zn, Pb, Cu, Cd)
- Chemical analysis of the humus layer (pH, element concentrations and stores, including trace elements) and mineral soil (pH, total C and N, cation exchange capacity *CEC* and its composition, element storage).

All measurements were conducted under the authority of the federal states and their own forestry research institutions, and chemical analyses were quality controlled by ongoing ring analyses (for further details see König and Wolff, 1993). The integration process was conducted by the *Federal Research Centre for Forestry and Forest Products (BFH)*, Eberswalde (see also Riek and Wolff, 1998).

Based on the integrated evaluation the following summary was given (BMELF, 1997):

- The data showed widespread acidification (see Fig. 2, distribution of pH-classes within the *BZE*-sample, pH measured in KCl and 0 - 10 cm soil depth) and loss of nutrient cations in the upper soil layers of almost all parent materials. Furthermore, there is an apparent tendency of increasing similarity in chemical conditions of the upper forest soil compartments on a low level of buffering capacity and fertility. Only sites with carbonates (limestone, marl) in the topsoil are not yet noticeably affected.

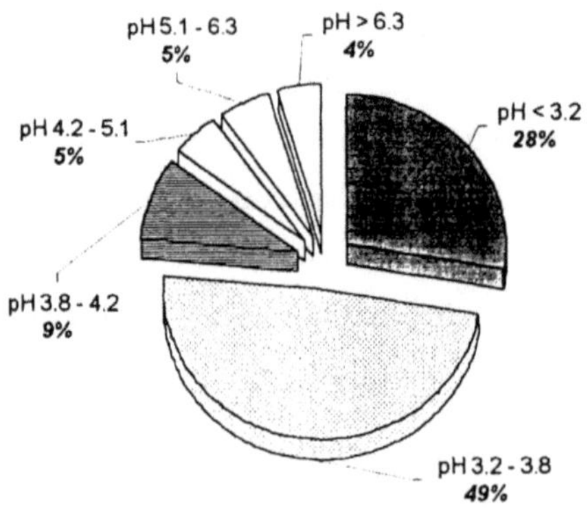

Figure 2. Percentage distribution of pH-classes within the BZE *soil survey sample, pH (KCl) in 0 - 10 cm soil depth (n = 1741, according to* BMELF, *1997)*

- The inconsistencies between pH values, C/N-ratios and humus morphology indicate that the forest soils are in a transformation process that is driven by atmospheric inputs of acidity and nitrogen compounds.
- From the deposition of strong mineral acids arises a significant risk for the quality of ground and spring water. Especially in podzolic, quartziferous soil types these acids can hardly be neutralised owing to low buffer capacities.
- The chemical strain imposed on forest sites by atmospheric pollution is also indicated by the heavy metal concentrations in the organic layers. In 25 % (resp. 38 %) of the survey plots, the amounts of Pb (resp. Cu) accumulated in the organic material are at concentrations that are potentially toxic to soil organisms. Considering the low pH-values in the topsoils, it must be assumed that significant parts of the more mobile heavy metals (Zn, Cd) have already percolated into deeper mineral layers.
- Especially for sites with low nutrient supply and unfavourable humus form, the biggest part of plant available nutrients was found in the organic layers, which can be considered as a merely metastable or even labile nutrient pool. In combination with simultaneous losses of base cations from the mineral layers, the affected forest ecosystems enter phases with increasingly unstable nutrient supplies.

- Regarding nutrients, the needle/leaf analyses of the tree species spruce (picea abies [L.] Karst.), pine (pinus sylvestris L.) and beech (fagus sylvatica L.), especially in the northern parts of Germany showed a luxuriant N supply and resulting imbalanced nutrient ratios (N/P, N/K, N/Mg). Low and extremely low P-concentrations are fairly common. Mg-deficiency symptoms mostly coincide with sandstone sites in the low mountain ranges.
- The S-concentrations in pine needles and beech leaves indicate a significant influence of atmospheric pollutants. 59 % of the analysed spruce stands show S-concentrations in needles that exceed the natural amounts and can therefore be regarded as affected by S-depositions.
- The examination of crown condition data led to the identification of two major groups of endangered sites. One of them includes sites with carbonaceous parent material, mostly of the rendzina soil type with a mull humus form. On these sites, forest damage symptoms in coniferous trees are increased. On the other hand, broad-leaved trees show damage on podzolic sites on pleistocene sands with a mor humus form.

For the general evaluation process the total sample of the survey was stratified by the variables *"parent material"*, *"soil type"*, *"humus form"*, *"elevation"*, and *"tree species"*. With respect to the first variable, about one third of the total survey sample originates from sites with quartziferous bedrock material (e.g. sites with eluvial or cover loam above sand-stone material, substrate group V, and poor sands, substrate group VI, see tab. 3). Within these two groups 19 to 31 % of sites are subject to a potential potassium deficiency (soil storage < 200 $kg \cdot ha^{-1}$, i.e. the total amount in the organic layer plus the exchangeable amount in the mineral soil down to a depth of 60 cm). 13 to 21 % of sites suffer from a potential magnesium deficiency (soil storage < 50 $kg \cdot ha^{-1}$, organic layer to 60 cm soil depth). For calcium only 4 to 7 % within the group of quartziferous substrates show poor storage values (soil storage < 200 $kg \cdot ha^{-1}$).

Other relevant substrate groups with higher percentages (12 - 18 %) of a potential potassium deficiency are various kinds of lime-free sediments (i.e. mainly sand loess and other loesses without lime admixture, and periglacial materials, substrate group III) as well as soils from substrate group IV (soils on eluvial or cover loam, clay-stone or silt rock material). These two substrate groups represent another third of the total survey sample. Relevant calcium storage deficiencies were found in substrate group IV, but also in soils originating from intermediate or acid magmatic and metamorphic substrates (mainly granites, gneisses). Those sites make up about 12 % of the total sample, the calcium deficiency being relevant for 10 % of this substrate group.

Considering different soil depths together with the above threshold values and elements, 40 % of all sites of the survey sample (i.e. % of all stratified soil substrates) suffer from a potassium deficiency, about 10 % from a magnesium-, and below 10 % from a calcium deficiency, when only a soil depth down to 30 cm is considered. Taking the 90 cm soil depth as the lower boundary, only 5 to 10 % of sites are in the range of a deficiency for all given elements. These findings illustrate that despite a strong upper soil acidification and a severe nutrient loss on some percent of the total forest area, most forest soils in Germany are sufficiently supplied with nutrient cations, when deeper soil horizons are considered.

	Substrate Groups	*[%] of total sample*	**Percentage of sites with *low** storage of** K	Ca	Mg
I	Calcareous loose sediments above carbonate rocks; carbonate bedrock material	*(9)*	1	0	0
II	Loose and carbonate-free sediments above carbonate bedrock material	*(13)*	3	3	5
III	Rearranged lime-free loose sediments	*(16)*	12	2	5
IV	Eluvial loam/cover loam above clay-stone/silt rock; clay, clay-stone, graywacke	*(13)*	18	9	3
V	Eluvial loam/cover loam above quartz-rich bedrock material	*(18)*	19	7	13
VI	Poor (pleistocenic) sands	*(15)*	31	4	21
VII	Intermediate/acid magmatic and metamorphic rocks, loam or mixed substrates above interm./acid magmatic and metamorphic rocks	*(12)*	5	10	2
	Others	*(4)*	-	-	-

Table 3. Stratified substrate groups of the BZE *soil survey, contribution of groups to the total sample and percentage with* low* *base cation storage within substrate groups (according to BMBF, 1997; n = 1124).* *(low base cation storage = < 200 kg K and Ca, < 50 kg Mg per ha, 0 to 60 cm soil depth, inclusive organic layer)

To assess more precisely the potential risk in forest management with special regard to nutrient cation supply and soil acidification, data from a strongly acidified, intensively monitored Norway spruce forest at the Solling, central Germany, will be presented as an example. The studied site represents a typical spruce monoculture plantation with an age of 67 years (in 2000) and is located on a plateau of the Solling mountains (51°31'N, 9°34'E, 500 m a.s.l; for more detailed site informations see Bredemeier et al., 1995a, for detailed forest yield data Dohrenbusch et al., 1993.). The soil is a spodic dystric Cambisol (FAO), developed from a solifluction of periglacial deposited loess (clay content about 15 %) and weathered sandstone parent material (clay content about 20 %). The mineralogy of the research area is characterised by a high content of Quartz (about 60 % of total grain size fraction in the upper 20 cm soil depth) and Illite (about 25 %) and relatively low contents of Feldspars (10 - 15 %) and Kaolinite (< 5 %, Bredemeier et al, 1995a). Texture is dominated by the silt grain size fraction. The humus layer is a mor with a thickness between 5 and 10 cm. The fine-root biomass amounts to a total of about 4 t·ha^{-1}, according to an inventory done in 1990 (Murach 1995). It is strongly concentrated in the humus layer (24 % of total) and the upper 30 cm of the mineral soil (65 % of total). Hence only about 10 % of the living fine-roots penetrate the mineral soil below a depth of 30 cm. Additionally, high amounts of dead fine-roots were found in the upper 30 cm (66 %, total 1.1 t · ha^{-1}) and the following 10 cm (21 % of total) of the mineral soil (Murach 1995). The reason for such a root architecture is probably the strong soil acidification and the low supply of base cations in the mineral soil.

Tab. 4 shows important indicators of soil chemical conditions at Solling. $pH_{(KCl)}$-values in the humus layer are below 3.0 and increase to only 4.2 in the 20 to 30 cm and deeper soil horizons. The exchange complex of the first 10 cm of the mineral soil is highly saturated with protons (19 % X^S = percentage of the total cation exchange capacity) and aluminium (70 %), whereas the saturation with base cations is low for all soil horizons. Below the 10 cm soil depth, aluminium is the dominant exchangeable cation. It is furthermore apparent, that the site exhibits relatively low C/N ratios for both horizons of the organic layer.

	$pH_{(KCl)}$	**C/N**	**C/P**		
Humus Layer					
L / Of	2.9	25	540		
Oh	2.7	25	382		
Mineral Soil		**Al^{3+}**	**H^+**	**(K^++ Ca^{2+}+ Mg^{2+})**	
			[X^S in %]		
0 - 10 cm	3.1	70	19	4	
10 - 20 cm	3.8	94	< 1	4	
20 - 30 cm	4.2	94	0	5	

Table 4. Soil chemical conditions for a Norway spruce (Picea abies (L.) Karst.) *plantation on loess over sandstone parent material in the Solling mountains, central Germany (*"Solling Roof Project", *pre-experimental conditions in 1988, see also Bredemeier et al., 1998 and Blanck et al., 1993).*

It seems that the humus morphology (here a mor type with a tendency to a raw humus) is not corresponding to the analysed C/N ratios. As already stated for the general soil survey, data from Solling corroborate a striking discrepancy between pH values, C/N-ratios and humus morphology, caused by acid atmospheric inputs as well as nitrogen deposition. Thus decaying plant residues are primarily accumulated in the organic layer and not mineralised and sequestered in the mineral soil, despite of relatively low C/N ratios.

Also base cations tend to accumulate most strongly in the organic layers of acidified forest soils. As an example it can be seen from fig. 3, that about 50 % (potassium) to more than 80 % (magnesium) of the total stores of base cations (here sum of organic layer to 80 cm soil depth) are accumulated in the organic horizon under spruce at the Solling. With respect to sudden mineralization pulses those reserves must be appraised to be highly labile. Natural disturbance events (e.g. wind fall, snow break) or forest management activities (e.g. thinning, skidding of logs) may increase light intensity and soil disturbance, both of which interferences promote mineralization of accumulated litter material.

It appears furthermore from fig. 3 that the total amount of base cations stored is low in comparison to the atmospheric input and the seepage output (here projected for a 90 year rotation period). About the same amount of potassium stored in the soil is deposited from the atmosphere within a 90 year rotation period, and about half of the atmospheric input is accumulated in the logs making up timber harvest (inclusive of the bark). The estimated weathering rate for potassium (212 kg per ha in 90 years) would not be enough to substitute the calculated accumulation in timber. For potassium as well as for calcium our flux budgets indicate negative values, i.e., there is a net loss of these elements during the calculation period. For magnesium a positive budget is calculated. This may be due to relatively high weathering rates (high clay content) and relatively low rates of accumulation in timber for magnesium, but soil reserves are extremely low (fig. 3).

The budgets compiled here are based on flux measurements for only five years (1990 - 1994), and estimates of weathering rates and accumulation rates in timber are subject to considerable uncertainties. Still, it is evident that strongly depleted forest sites of the Solling spruce type are highly dependent on atmospheric base cation inputs, the budget also depending strongly on seepage output fluxes. Changes in flux rates could easily impair the nutrient supply, due to the relatively low storage potential and low amounts of nutrients actually stored.

From a soil chemical and microbiological point of view, an active basicitiy management of forest soils has to be strongly encouraged. Accumulated humus material and litter with a positive alkalinity is a potential source of internal basicity generation and base cation input, but decomposition and turnover rates have to be stimulated, before base saturation in soils will increase. This stimulation can be achieved by external addition of bases in the form of lime or wood ash, such that internal generation of basicity may be triggered by external addition. Buffer range and pH of forest soils will only improve very slowly, even under reduced present atmospheric deposition, unless actual basicity is added to the soil system.

	$K_{(TD)}$	$Ca_{(TD)}$	$Mg_{(TD)}$
Atmospheric input	+ 471	+ 577	+ 203
Accumulation in stem timber (incl. bark)	- 260	- 100	- 30
Weathering rate (0 - 90 cm soil depth)	+ 212	+ 108	+ 361
Output (100 cm soil depth)	- 607	- 714	- 337
Budget	**- 184**	**- 129**	**+ 197**
Soil storage			
Org. layer	215	163	77
Mineral soil *(0 – 80 cm soil depth)*	246	**70**	**18**
Sum	**461**	**233**	**95**

Figure 3. Element flux budgets and stores for a Norway Spruce forest at Solling, central Germany. Flux budgets are projected estimates for a 90-year rotation period [kg ha^{-1}]. TD = *total deposition (bulk precipitation + interception deposition, calculated according to Ulrich, 1983) and output are based on 1990 to 1994 measurements for the control plot D0 of the* "Solling Roof Project" *(see Xu et al., 1998); above ground biomass according to Dohrenbusch et al. (1993), weathering rates from Wesselink et al. (1994), soil stores according to Blanck et al. (1993).*

5 Silvicultural strategies to improve site conditions and reduce risk in German state forest management

Silvicultural strategies to reduce risk are essentially aimed at an improvement of the forests' capability of adaptation to changing environmental conditions. Today most foresters share the conviction that the particularly high risks arising from the long production periods in forestry can be best faced by creating forests that are diverse and resemble potential natural states. Modern silviculture seeks to understand and then to mimic the strategies of nature in achieving persistence of forest ecosystems, and to make use of such strategies for the aims of forest management. In the following passages, we discuss major fields of silvicultural decisions under this aspect.

Tree species selection

The many risk factors constituting a probability of damage (p_d, see above) that can not be calculated a priori with certainty are demanding a tree species selection strategy which is not only oriented at tolerance against one single stress factor. Particular tolerance against one single factor often goes along with increased sensitivity against others. For example, in conifers a generally higher ozone tolerance compared to deciduous species seems to be connected with a higher sensitivity to SO_2 (Guderian and Wienhaus 1997).

An active adaptation of siliviculture to a future climate in Europe is hardly possible, since that future climate constellation can not be known with adequate certainty. The same situation is given with respect to the rise in atmospheric CO_2 concentration. Although the latter is a well documented and quantified environmental change (in contrast to the actual und future extent of climatic change), the responses of tree species and forests to this growth factor are as yet poorly known.

Under these uncertainties minimization of risk can only be anticipated if tree species (or provenances, i.e. populations within species) are selected for a high general stress tolerance. This reasoning is not new in forestry (Röhrig and Gussone 1990), however, it is gaining more weight presently. Many experiences (e.g. in areas of smoke damage) support the conclusion that tree species, whose autecological requirements are optimally met at a site, are more resistant against stresses (Wentzel 1964). This conclusion is also supported by a comprehensive evaluation of the more recent literature on forest decline, particularly with respect to water availability (Innes 1993). Concerning the potential future climatic change, Thomasius (1997) presumes that species at the limits of their ecological amplitudes will be more endangered than those inhabiting sites within their optimum range. Hence he suggests to integrate species with a wide site amplitude (e.g. the pioneers birch and pine) more strongly into silvicultural planning. However, possibilities to do so are limited, given the still greatly important goal of wood production. The focal point of species selection in German forests will most probably remain with the economically most important species spruce (*Picea abies*), beech (*Fagus sylvatica*) and oaks (*Quercus robur* and *Q. petraea*). They have a narrower site amplitude and should therefore only be selected in the future for sites corresponding closely to their respective optima.

The cultivation of spruce entails special problems. Norway spruce is the most economically important species in Central Europe; at the same time, it is the species most subject to drought, storms, snow and insects (see Schmidt-Vogt 1991 for a comprehensive summary), and it exerts the strongest detrimental effects on site quality (Rehfuess 1990). The further cultivation of spruce is therefore subject to an intense dispute. In the study cited above, Schmidt-Vogt has thoroughly analyzed all aspects of spruce cultivation, and concludes that spruce will continue to play a major role in European silviculture, provided that it is cultivated at suitable sites, and that stand densities are strongly reduced at an early stage. It is crucial for site selection that the water requirements of Norway spruce can be met, and that deep rooting is feasible for the trees. A prerequisite to achieve the latter are well aerated and sufficiently buffered soils (Puhe 1994).

The selection of autochthonous species and populations can no more guarantee best adaptation to the site and abatement of risk when key environmental variables are changing. Non-autochthonous populations or species (neophytes) may exhibit better stress tolerance under such constellations. Examples are white fir (*Abies alba*) populations from Calabria, southern Italy, that proved more vital in a cultivation trial in Denmark than Central European provenances (Larsen 1981), or Douglas fir plantations (*Pseudotsuga menziesii*), that exhibit a large site amplitude and drought tolerance in Central Europe (Röhrig and Bartsch 1992, Otto 1993 b).

Forest Ecosystems of high Diversity

A further means to indirectly increase stress tolerance and decrease risk is the establishment of forest ecosystems with a high degree of diversity. The reason is that the constituents (tree populations or individuals) in such ecosystems have different reaction patterns to stress factors. A certain stress or risk factor is thus only acting upon part of the constituents of the system. Hence, the risk of strong damages is distributed, and it can be expected that for the ecosystem as a whole risk will decrease.

Owing to its history with massive afforestations of devastated forest areas in the last century, central European forestry has extensive experience with the specific risks of large-scale, homogeneous monocultures, particularly of Norway spruce. As a consequence of this experience, all forest administrations of the federal states in Germany give an almost exclusive preference to mixed stands in their silvicultural strategies for forest regeneration. Otto (1986) gives a broad reasoning for this approach with respect to the Harz mountain region in Lower Saxony.

In this region, like in many other forest areas of Central Europe, the focus of forest planning and management today is on transformation of conifer monocultures into mixed stands with broadleaf tree, particularly European beech. The impetus for this decision arose from the damages observed in spruce crowns, which were attributed to the emission of air pollutants and acid deposition. It is expected that mixed stands will have an improved tolerance to emissions, due to the higher SO_2 tolerance of broadleaf trees, together with a lower filtering efficiency and hence lower deposition rates (summarized in Röhrig and Bartsch 1992, Beese 1997). Further advantages are expected with respect to an improvement of general vitality: the changed composition of litter and the lowered acid deposition under a canopy with broadleafs should improve decomposition conditions (Brown, 1992; Lücke and Schmidt, 1997; Mettin, 1985), and thus accelerate nutrient cycling. Since broadleaf trees usually exhibit distinctly deeper rooting than Norway spruce, nutrient reservoirs in the deeper soil could be accessed (Block et al. 1997). However, Rothe (1997) points out that these effects could only be observed with beech admixtures > 40 % in his studies.

Many forest practitioners expected a higher stability of Norway spruce against windthrow when spruce grows in mixed stands. However, in the analyses of damage after the heavy 1990 storms, no consistent pattern of such kind could be found (Lüpke and Spellmann, 1997). Nevertheless, under heavy storms the physically more stable tree species will survive with a higher probability, so the spatial extent of damage will be limited, and regeneration of the stand will be facilitated. A higher resilience of the forest can also be expected regarding insect damages, which normally selectively hit one tree species. For the same reason, an improved resilience of mixed forest to the effects of a possible climatic change can be anticipated.

Similar to the differences in tolerance between species, the different age stages in one species may respond differentially to stresses. Top thinning (needle loss), for instance, is typical for old Norway spruce stands, whereas younger stands are more susceptible to snowbreak. Consequently, there is a gain of stability due to enhanced resilience even in monospecies stands, if trees of different ages are closely mixed. High impacts of a selectively acting stressor can only damage parts of the entire stand area in that situation, regeneration after disturbance can proceed quicker, based on the undamaged trees, than after a large-scale destruction. Forests with a well developed age structure are therefore generally considered as a suitable means in silvicultural management to strengthen stability (Otto, 1993 a).

At the level of populations genetic diversity is determining the capability of adaptation to changing environmental conditions. This feature is especially important in long-lived organisms such as trees (Müller-Starck, 1993; Degen, 1997). In times of rapidly changing environmental conditions it is therefore imperative as a forest management goal to achieve a maximum conservation of genetic variability (Larsen, 1996). Beyond in-situ or ex-situ gene conservation programmes, this goal has to be pursued on the entire forest area by employing appropriate practical silivicultural measures. Degen (1997) and Schoppa (1999) give some concrete indications of such measures. According to them, natural forest regeneration should be preferred to sowing or planting wherever possible. A large population efficient in reproduction is crucial for conservation of genetic variation. Under a comparable degree of genotypical variation, populations with many individuals have a better potential of adaptation than those with few individuals. Hence, in this respect a dense natural rejuvenation with its countless individuals is optimal, and a large population resulting from sowing is better than a small one from planting. In the latter category, narrow spaced plantations would be more favourable than wide spaced ones with only few individuals.

If tree populations are selected by intensive breeding for optimal performance in one particular respect, e.g. volume increment or SO_2 resistance, their genetic diversity decreases, with the likely consequence of a diminished capability for adaptation. As a result, this method to reduce single risk factors while neglecting others is not favoured any more. Instead, Karnosky and Scholz (1996) suggest the following two aims for forest tree breeding: (1) general enhancement of genetic diversity at the population level and (2) selection of individuals with an apparently high and broad stress tolerance. However, no practical applications of this theoretical concept are known as yet.

Minimal disturbance by forest utilization operations

All utilization of forests comprising biomass export also means an export of nutrients from the ecosystem. At sites with poor soils, this may lead to an intolerable increase in risk (Ulrich 1995, see also sections above). Under such circumstances forest utilization has to be strictly limited to bolewood only, whole-tree harvesting must be avoided. The slash (twigs, branches, bark) should remain in the forest. It should neither be concentrated in piles or walls on the site, nor should it be burned. The length of the rotation is important as well: long rotations entail a less frequent export of big logs, which removes less nutrients than a frequent export of smaller logs. Overall nutrient exports are higher with short rotations.

The felling of a tree disrupts nutrient uptake at that location, if its soil volume is not exploited by neighbouring trees. If only single trees are harvested (e.g. in the case of a thinning), continuing root uptake by neighbours is probably often given. If a number of neighbouring trees are cut, however, root uptake at that location ceases (from about 10-15 m diameter of a gap cut onwards). Since litter decay and mineralization of nutrients continue, possibly accelerated by the changed microclimatic conditions, nutrient losses with seepage water can be expected (Bartsch et al., 1999; Bauhus and Bartsch, 1995). Figures 4 and 5 show examples of nitrate losses and calcium losses, respectively.

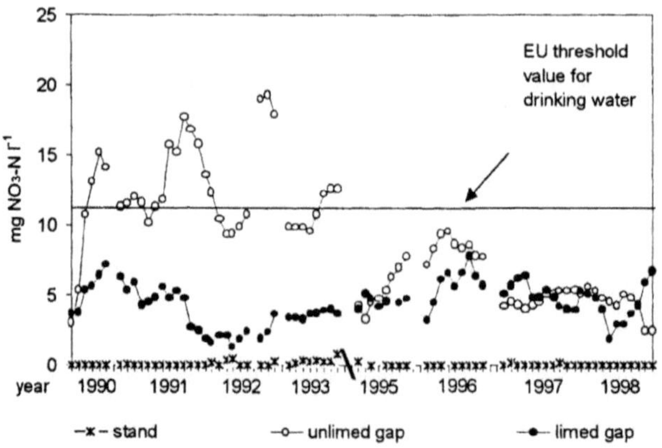

Figure 4: Time courses of nitrate-N concentrations in seepage water (80 cm soil depth) in a gap cut experiment at Solling (after Bartsch et al., 1999)

This is one of the reasons why all forest administrations of the federal states in Germany now refrain from large-scale clearcutting. Other reasons are the advantages of alternative methods in establishing highly structured forest stands, as well as economic and conservation considerations. Harvest operations are done on small spatial extensions, and only when a rejuvenation is already established under the shelter of the old growth. Nutrient losses and hazards to water quality are minimized in this way. A problem with this management scheme that avoids clearcuts arises when less shade tolerant species are to be regenerated, e.g. the oaks, pine or European larch *Larix decidua* (Lüpke, 1992). Larger open areas for the cultivation of these species frequently result from windthrow, but it is naturally the intention of the forest managers to avoid large windthrows.

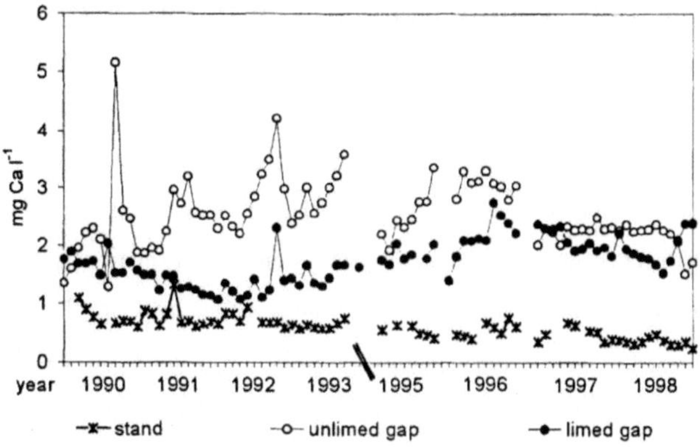

Figure 5: Time courses of calcium concentrations in seepage water (80 cm soil depth) in a gap cut experiment at Solling (after Bartsch et al., 1999)

The quality of the soil as a rooting medium is also imperiled by driving over the surface with heavy machinery (see the review by Schack-Kirchner 1996). Bulk density is increased, porosity decreased and the transport of soil water and air impeded. This gives rise to additional stress, which may act particularly detrimentally in extreme situations like drought or acute nutrient deficiency; the regeneration in forests seems to be especially sensitive. Such detrimental effects can be minimized by a strict concentration and limitation of all machinery movement to a few well defined skidder lines.

Tending of stands

It is the aim of thinnings to assist selected trees against competition, thus improving their vitality and increment. The expectation is that these supported trees, with their longer and bigger crowns and larger rooting spaces, will develop a better stability against storms, and an improved tolerance against risk factors in general. At the same time, every thinning is an interference with the neighbourhood structures in a forest stand, which may bring about stress, until the remaining trees have adapted to the new microclimatic conditions and have established a new structural matrix. This source of stress can be reduced if thinnings are done with lower intensities, but with a higher frequency, beginning at early stand ages. Breaking up closed structures in an old growth stand always bears considerable risks. Also, structural disruptions caused by bark beetles and other insects should be avoided by adhering to principles of clean management (i.e., to remove material suitable for insect breeding in due time).

6 Conclusions: environmental changes and hazard rating

Present environmental changes affecting forest ecosystems reflect a long history of change and human interference. We outlined how with human settlement and exploitation of the landscape, site conditions and ecosystem structure in forests were seriously modified and frequently deteriorated to the point of severe degradation. The spectrum of environmental changes, human interference and the subsequent ecosystem transformations are multitudinous and operate on different spatial and temporal scales, depending on the processes and responses under consideration. For European forest soils it was demonstrated that numerous human activities and exploitation practices contributed to a deterioration of site quality, and that the period of interference comprises more than a millenium. Long before such modern phenomena as acid rain or rising CO_2 levels, forest ecosystems were seriously affected by natural environmental fluctuations (e.g. the "little ice age") and human interference.

A particular quality of recent environmental changes, however, is that they probably occur much more rapidly than natural oscillations in climate, and even more, that they lead to constellations of ecological factors that may be unprecedented in history and evolution. As we showed, peculiar combinations of growth and vitality factors arise from recent environmental changes ("recent" is to be understood here as pertaining to the period of industrialization, i.e. ca. 150 years). Some factors under change have the potential to foster vitality and growth, e.g. CO_2 enrichment and enhanced nitrogen deposition. Others may have strong detrimental effects, like soil acidification or compaction. It is a major task of ongoing research, e.g. (Rehfuess et al., 1999) to resolve and quantify the interactions of the various current environmental changes – a task that has only just commenced and is far from being completed.

For forest management it is crucially important to have good information on site properties and site quality. This pertains to the stand level as well as to the forest district and enterprise, the landscape- and the regional level up to national level. We reported on nationwide investigation of forest soil status in Germany, revealing that the majority of forest soils are indeed affected by acidification, at least in the upper soil layers that are most important for rooting. Soil amelioration with inputs of basicity (lime, wood ash) may reduce hazards resulting from soil acidity, but are constrained by technical and economical feasability.

Adapted silvicultural strategies have the potential to significantly reduce risk with respect to many of the contributing factors. More diverse forest ecosystems mimicking structures and functions in natural forests are expected to bring about more stability. However, economic constraints may put limits on this development, at least in the private sector. Still, in Germany public forest administrations have decided to adopt more ecologically oriented silvicultural practices. If these prove to also be economically viable due to their better control of risk, they may be adopted by the private forestry sector as well.

Our argument is that risk in practical forestry can hardly be quantified using the classical risk formula, which calculates risk as the numerical product of probability and amount of damage. For such a quantification, information on the probability of damage is simply lacking, at least at the forestry management unit level, the individual stand or site. Nevertheless, qualitative assessments can be performed with some confidence, and they should be performed before making management decisions, that will influence forest structure and function for decades. An example of such a qualitative assessment scheme is given in Tab. 5 in the form of a cross-tabulation of some most important environmental risk factors and Central European tree species.

Risk type	*Picea abies*	*Pinus sylvestris*	*Fagus sylvatica*	*Quercus petrea*	*Quercus robur*	*Fraxinus excelsior*	*Acer pseudopl.*
Windthrow	+++++	+++	++	+	+	++	++
Drought	+++++	+	+++	+	+	++	+++
Frost	+	+	+++	++++	+++	+++	++
Waterlogging	++++	++	+++++	++	+	++	++++
Soil acidity	++	+	+	+	+	+++++	+++++
Insects	++++	++++	+	++++	++++	++	++

Table 5. Example of qualitiative risk ranking for seven Central European forest tree species and six major risk categories; number of symbols in the cross-tabulation cells indicates relative risk (+ to +++++).

Depending on how much information is available on the potential weight of individual risk factors at an individual site, such risk assessments should be able to help prepare and improve forest management decisions.

Acknowledgement

Projects contributing to this paper were funded by the German Federal Ministery of Education and Research (BMBF) under funding no. 0339474 and the Federal State of Lower Saxony.

References

Aber, J.D., Nadelhoffer, K.J., Steudler, P.A. & Melillo, J.M., 1989: Nitrogen saturation in Northern forest ecosystems. *Bioscience* **39**, 378-386.

Banse, G., 1996: Herkunft und Anspruch der Risikoforschung. In: Banse, G., (Ed.) *Risikoforschung zwischen Disziplinarität und Interdisziplinarität*, pp. 15-27. Berlin.

Bartsch, N. & Bauhus, J., Vor, T., 1999: Auswirkungen von Auflichtung und Kalkung auf das Sickerwasser in einem Buchenbestand (Fagus sylvatica L.) im Solling. Forstarchiv **70**, 218-223.

Bauhus, J. & Bartsch, N., 1995: Mechanisms for carbon and nutrient release and retention in beech forest gaps. I. Microclimate, water balance and seepage water chemistry. Plant and Soil. **168-169**: 579-584.

Beese, F. 1997: Wirkungskomplex Waldboden und Waldgesundheit. In: Deutscher Forstverein e.V. (ed.): Deutscher Forstverein, Jahresbericht 1996, Niedenstein, 201-206.

Beese, F.O. 1996: Indikatoren für eine multifunktionelle Waldnutzung. *Forstw.Cbl.* **115**, 65-79.

Binkley, D. & Hogberg, P., 1997: Does atmospheric deposition of nitrogen threaten Swedish forests [Review]. *Forest Ecology & Management* **92**, 119-152.

Blanck, K., Lamersdorf, N. & Bredemeier, M., 1993: Bodenchemie und Stoffhaushalt auf den Dachflächen im Solling. *Forstarchiv* **64**, 164-172.

Block, J., Roeder, A. & Schüler, G., (1997): Waldbodenrestauration durch Aktivierung ökosystemarer Nährstoffkreisläufe. AFZ/DerWald. **52**: 29-33.

BMELF (Bundesministerium für Ernährung, Landwirtschaft und Forsten) 1994: Bundesweite Bodenzustandserhebung im Wald (BZE), Arbeitsanleitung (2. Ed.), Bonn.

BMELF (Bundesministerium für Ernährung, Landwirtschaft und Forsten) 1997: Deutscher Waldbodenbericht 1996 (Bd. 1), Bonn.

Borken, W. & Brumme, R., 1997: Liming practise in temperate forest ecosystems and the effects on CO_2, N_2O and CH_4 fluxes. Soil Use and Management, **13**, 251-257.

Braess, P. 1960: *Versicherung und Risiko*, Wiesbaden: Gabler.

Bredemeier, M., Blanck, K. & Wiedey, G.A., 1993: Experimentelle Manipulation des Wasser- und Stoffhaushalts in einem Fichtenwald - das Dachprojekt im Solling. *Forstarchiv* **64**, 154-158.

Bredemeier, M., Tiktak, A., & van Heerden, K., 1995: The Solling Norway Spruce site. *Ecological Modelling* **83**, 7-15.

Bredemeier, M., Dohrenbusch, A. & Murach, D., 1995a: Response of soil water chemistry and fine-roots to clean rain in a spruce forest ecosystem at Solling, FRG. *Water, Air, & Soil Pollution* **85**, 1605-1611.

Bredemeier, M. & Fölster, H., 1998: Dynamics, stability and scales in forest ecosystems. *Senckenbergiana maritima* **27**, 245-249.

Bredemeier, M., Blanck, K., Dohrenbusch, A., Lamersdorf, N., Meyer, A.C., Murach, D., Parth, A., & Xu, Y.J., 1998: The Solling roof project - site characteristics, experiments and results. *Forest Ecology and Management* **101**, 281-293.

Brown, A.H.F., 1992: Functioning of mixed-species stands at Gisburn, N.W. England. In: Cannell, M.G.R.; Malcolm, D.C.; Robertson, P.A. (ed.): The ecology of mixed-species stands of trees. Special publication series of the British Ecological Society, **11**, Blackwell Scientific Publications, London et al., 125-150.

Brumme, R., Borken, W. & Finke, S., 1999: Hierarchical control on nitrous oxide emission in forest ecosystems. *Global Biogeochemical Cycles* 1137-1148.

Degen, B., 1997: Ökologische Genetik - Abschlußbericht -. In: Umweltbundesamt (ed.): Auswertung der Waldschadensforschungsergebnisse (1982-1992) zur Aufklärung komplexer Ursache-Wirkungsbeziehungen mit Hilfe systemanalytischer Methoden. Berichte, **6/97**, Erich Schmidt Verlag, Berlin, 407-521.

Dohrenbusch, A., Grote, R. & Fritz, H.-W., 1993: Struktur und Wachstum eines Fichtenbestandes unter experimenteller Manipulation der Stoffeinträge. *Forstarchiv* **64**, 172-177.

Ellenberg, H., 1988: *Vegetation Ecology of Central Europe*, 4 edn. Cambridge: Cambridge Univ. Press.

Emmer, I.M., Wessel, W.W., Kooijman, A., Sevink, J. & Fanta, J., 2000: Restoration of degraded Central-European mountain forest soils under changing environmental conditions. *Plant and Soil* **in press**,

Falkengren-Grerup, U. & Tyler, G., 1992: Changes since 1950 of mineral pools in the upper C-horizon of Swedish deciduous forest soils. *Water, Air, & Soil Pollution* **64**, 495-501.

FAO (Food and Agriculture Organization of the United Nations) 1995: FAOSTAT-PC, Land use. Rome, Italy.

Fölster, H., 1992: Holocene autochthonous forest degradation in Southeast Venezuela. In: Goldammer, J.G., (Ed.) *Tropical Forests in Transition*, pp. 25-44. Basel: Birkhäuser

Godbold, D.L., 1994: Aluminium and heavy metal stress: From the rhizosphere to the whole plant. In: Godbold, D.L. and Hüttermann, A. [Eds.] Effects of acid rain of forest processes, pp. 231-264. New York. Wiley-Liss.

Godbold, D.L., 1998: Stress concepts and forest trees. *Chemosphere* **36**, 859-864.

Godbold, D.L. & Jentschke, G., 1998: Aluminium accumulation in root cell walls coincides with inhibition of root growth but not with inhibition of magnesium uptake in Norway Spruce. *Physiologia Plantarum* **102**, 553-560.

Guderian, R. & Wienhaus, O., 1997: "Neuartige Waldschäden" und Luftverunreinigungen aus gegenwärtiger Sicht. In: Deutscher Forstverein (ed.): Deutscher Forstverein, Jahresbericht 1996, Niedenstein, 181-200.

Innes, J.L., 1993: Forest health: its assessment and status. CAB International, Wallingford (UK), 677.

Karnosky, D.F. & Scholz, F., 1996: Managing the genetic structure of stressed forests. Congress report, IUFRO XX World Congress 1995, Tampere, Finland, 362-368.

König, N. & Wolff, B., 1993: Abschlußbericht über die Ergebnisse und Konsequenzen der im Rahmen der bundesweiten Bodenzustandserhebung im Wald (BZE) durchgeführten Ringanalysen. *Berichte des Forschungszentrums Waldökosysteme Göttingen*, Reihe B, **Bd. 33**, 1-33.

Kuhnt, G., Garniel, A., Kothe, P., & Schröder, W., 1991: Standortbestimmung für die begleitende Bodenprobennahme und -analyse sowie Überprüfung der Meßnetzvalidität. *BMU Forschungsbericht* 107 06 002.

Lamersdorf, N., 1988: Verteilung und Akkumulation von Spurenstoffen in Waldökosystemen (Diss.). *Berichte des Forschungszentrums Waldökosysteme Göttingen,* Reihe A, **Bd. 36**.

Larsen, J.B., 1981: Waldbauliche und ertragskundliche Erfahrungen mit verschiedenen Provenienzen der Weißtanne in Dänemark. Forstwiss. Cbl. **100**: 275-286.

Larsen, J.B., 1996: Silviculture and the stability of stressed forests. Congress report, IUFRO XX World Congress 1995, Tampere, Finland, 343-351.

Lücke, K. & Schmidt, W., 1997: Vegetation und Standortsverhältnisse in Buchen-Fichten-Mischbeständen des Sollings. Forstarchiv **68**: 135-143.

Lüpke, B. v., 1992: Waldbau ohne Kahlschlag - Möglichkeiten und Risiken. Forstarchiv. **63**: 10-15.

Lüpke, B. v. & Spellmann, H., 1997: Aspekte der Stabilität und des Wachstums von Mischbeständen aus Fichte und Buche als Grundlage für waldbauliche Entscheidungen. Forstarchiv **68**: 167-179.

Matzner, E. & Murach, D., 1995: Soil changes induced by air pollutant deposition and their implication for forests in central Europe. Water, Air, & Soil Pollution **85**, 63-76.

Meesenburg, H., Meiwes, K.J. & Rademacher, P., 1995: Long-term trends in atmospheric deposition and seepage output in northwest German forest ecosystems. Water, Air, & Soil Pollution **85**, 611-616.

Mettin, C., 1985: Betriebswirtschaftliche und ökologische Zusammenhänge zwischen Standortskraft und Leistung in Fichten-Reinbeständen und Fichten/Buchen-Mischbeständen. AFZ **40**: 803-810.

Möhring, C., 1992: 10 Jahre Waldschadensforschung - Bilanz und Ausblick. pp.3-72. Bonn: Bundesminister für Forschung und Technologie.

Müller-Starck, G., 1993: Auswirkungen von Umweltbelastungen auf genetische Strukturen von Waldbeständen am Beispiel der Buche (Fagus sylvatica L.). Schriften a.d.Forstl.Fak.d.Univ.Göttingen u.d.Nds.Forstl.Vers.Anst., **112**, Sauerländer's Verlag, Frankfurt/M., 163 S.

Murach, D., 1995: Fine root investigations in the Solling Roof Experiment. In: Jenkins, A., Ferrier, R.C. and Kirby, C. (Eds.), ecosystem manipulation experiments: Scientific approaches, experimental design and relevant results. *Commission of the European Communities, Ecosystems Research Report* **20**, 155-158.

Oldeman, L.R., 1992: Global extent of soil degradation. In: Anonymous *ISRIC, Bi-annual report 1990-1992*, pp. 19-39. Wageningen, The Netherlands: ISRIC.

Ortloff, W. & Schlaepfer, R., 1996: Nitrogen deposition and forest health - a literature review [Review] [German]. *Allgemeine Forst und Jagdzeitung* **167**, 184-201.

Otto, H.-J., 1986: Standörtliche Voraussetzungen, Ziele und Waldbautechnik in Fichten-Buchen-Mischbeständen des Harzes. Teil I: Standorte, Leistungsstruktur und Geschichte der Mischbestände von Fichte und Buche im Harz. Teil II: Ziele, Möglichkeiten und Waldbautechnik von Mischbeständen der Fichte und Buche im Harz. Allg. Forst-u. Jagdzeitung. **157**: 188-196, 214-222.

Otto, H.-J., 1993 a: Der dynamische Wald. Ökologische Grundlagen des naturnahen Waldbaues. Forst u. Holz **48**: 331-335.

Otto, H.-J., 1993 b: Fremdländische Baumarten in der Waldbauplanung - dargestellt am Beispiel der niedersächsischen Landesforstverwaltung -. Forst u. Holz **48**: 454-456.

Papen, H. & Butterbach-Bahl, K., 1999: A 3-year continuous record of nitrogen trace gas fluxes from untreated and limed soil of a N-saturated spruce and beech forest ecosystem in Germany - 1. N2O emissions. *Journal of Geophysical Research-Atmospheres* 18487-18503.

Pearson, J. & Soares, A., 1995: A Hypothesis of Plant Susceptibility to Atmospheric Pollution Based on Intrinsic Nitrogen Metabolism - Why Acidity Really is the Problem. Water, Air, & Soil Pollution **85**, 1227-1232.

Persson, H. & Majdi, H., 1995: Effects of Acid Deposition on Tree Roots in Swedish Forest Stands. Water, Air, & Soil Pollution **85**, 1287-1292.

Plochmann, R., 1979: Mensch und Wald. In: Stern, H., (Ed.) *Rettet den Wald*, pp. 157-197. München, Kindler

Puhe, J., 1994: Die Wurzelentwicklung der Fichte (Picea abies [L.] Karst.) bei unterschiedlichen chemischen Bodenbedingungen. Berichte d. Forschungszentrums Waldökosyst., Reihe A, **108**, Göttingen, 128 S.

Rehfuess, K.E., 1990: Waldböden. Verlag Paul Parey, Hamburg u. Berlin, 2. Auflage, 224 S.

Rehfuess, K.E., Agren, G.I., Andersson, F., Cannell, M., Friend, A., Hunter, I., Kahle, H.-P., Prietzel, J. & Spiecker, H., 1999: Relationships between recent changes of growth and nutrition of Norway spruce, Scots pine, and European beech forests in Europe. EFI Research Report **19**, pp.1-94. European Forest Institute.

Rennenberg, H., Kreutzer, K., Papen, H. & Weber, P., 1998: Consequences of high loads of nitrogen for spruce (Picea abies) and beech (Fagus sylvatica) forests. *New Phytologist* **139**, 71-86.

Richter, C.M., Kranig, S. & Wild, A., 1995: Contents of Free Amino Acids in Needles of Norway Spruce Trees in Relation to Novel Forest Decline - Studies on Trees From a Site in the Northern Black Forest. Environmental Pollution **87**, 303-312.

Riek, W. & Wolff, B., 1998: Integrierende Auswertung bundesweiter Boden-, Kronenzustands- und

Baumernährungsdaten. *AFZ/DerWald* **10**, 511 - 513.

Röhrig, E. & Bartsch, N., 1992: Der Wald als Vegetationsform und seine Bedeutung für den Menschen. 6. Auflage von DENGLERS "Waldbau auf ökologischer Grundlage", Band 1, Paul Parey, Hamburg und Berlin, 350 S.

Röhrig, E. & Gussone, H.A., 1990: Baumartenwahl, Bestandesbegründung und Bestandespflege. 6. Auflage von DENGLERS "Waldbau auf ökologischer Grundlage", Band 2, Paul Parey, Hamburg und Berlin, 314 S.

Rothe, A., 1997: Einfluß des Baumartenanteils auf Durchwurzelung, Wasserhaushalt, Stoffhaushalt und Zuwachsleistung eines Fichten-Buchen-Mischbestandes am Standort Höglwald. Forstl. Forschungsber. München, **163**, 213 S.

Schack-Kirchner, H., 1996: Holzernte und Bodenschutz - Wie ernst muß man das Vorsorgeprinzip nehmen?. In: Waldwirtschaft und Waldökologie. Agrarforschung in Baden-Württemberg, **26**, Eugen Ulmer Verlag, Stuttgart, 259-267.

Schmidt-Vogt, H., 1991: Die Fichte. Bd. **II/3**, Paul Parey, Hamburg u. Berlin, 781 S.

Schmieden, U. & Wild, A., 1995: The contribution of ozone to forest decline [Review]. *Physiologia Plantarum* **94**, 371-378.

Schober, R., 1975: *Ertragstafeln wichtiger Baumarten*, Frankfurt: Sauerländer.

Schoppa, F., 1999: Konsequenzen wald- und forstgeschichtlicher Entwicklungen für die aktuelle genetische Zusammensetzung von Waldbaumpopulationen in Mitteleuropa. Diss. Fak. f. Forstwiss. u. Waldökologie d. Uni. Göttingen, 162 S.

Schroeder, J.v. & Reuss, C., 1883: *Die Beschaedigung der Vegetation durch Rauch und die Oberharzer Huettenrauchschaeden*, Berlin.

Schweiger, S., 1989: Vergleich dreier bewaldeter Einzugsgebiete (Hoylandet, Birkenens, Lange Bramke) entlang eines Nord-Süd-Gradienten der Säuredeposition anhand ausgewählter Ökosystemparameter - Bodenchemische Aspekte. Diplomarbeit, Geogr. Inst. der Univ. Göttingen.

Spiecker, H., 1999: Overview of recent growth trends in European forests. *Water, Air, & Soil Pollution* 33-46.

Svolba, J. (1997): Entwicklung eines Herkunftsversuches der Weißtanne (Abies alba MILL.) bis zum Alter 27. In: Gagov, V. (ed.): Ergebnisse des 8. Tannen-Symposiums. Schriften d. IUFRO u.d. Forsttechn. Univ. Sofia, Sofia, 77-88.

Thomasius, H., 1997: Waldbau im Zeichen starker Umweltveränderungen. In: Deutscher Forstverein e.V. (ed.): Jahresbericht 1996, Niedenstein, 207-228.

Ulrich, B., 1972: Chemische Wechselwirkungen zwischen Waldökosystemen und ihrer Umwelt. *Forstarchiv* **43**, 41-43.

Ulrich, B., 1981: Theoretische Betrachtung des Ionenkreislaufs in Waldökosystemen. *Z.Pflanzenernähr.Bodenkde.* **144**, 647-659.

Ulrich, B., 1983: A concept of forest ecosystem stability and of acid deposition as driving force for destabilization. In: Ulrich, B. and Pankrath, J., (Eds.) *Effects of accumulation of air pollutants in forest ecosystems*, pp. 1-29. Dordrecht: D. Reidel

Ulrich, B., 1994: Process hierarchy in forest ecosystems: an integrating ecosystem theory. In: Godbold, D.L. and Hüttermann, A., (Eds.) pp. 353-397. New York: Wiley-Liss]

Ulrich, B., 1994: Nutrient and acid-base budget of central European forest ecosystems. In: Godbold, D. L. and Hüttermann, A. (Eds.), effects of acid rain on forest processes. *Wiley-Liss*, Inc., 1-50.

Ulrich, B., 1995: Der ökologische Bodenzustand - seine Veränderung in der Nacheiszeit, Ansprüche der Baumarten. Forstarchiv **66**: 117-127.

Umweltbundesamt, 1997: Daten zur Umwelt. Der Zustand der Umwelt in Deutschland. Ausgabe 1997, 6th ed. Berlin: Erich Schmidt Verlag.

Various Authors, 1993: *Restoration of tropical forest ecosystems : proceedings of the symposium held on October 7-10, 1991*, Dordrecht: Kluwer.

WBGU, 1994: World in transition: the threat to soils. 1994 Annual Report, pp.1-252. Bonn: Economica.

Wentzel, K.F., 1964: Gibt es immissionsfeste oder rauchharte Bäume? Forstarchiv **35**: 49-51.

Wesselink, B., van Grinsven, H. & Großkurth G., 1994: Mechanisms and rates of silicate weathering at Solling, Germany. In: Wesselink, L.G., Time trends & mechanisms of soil acidification. *PhD thesis Wageningen*, 29-52.

Whytemare, A.B., Edmonds, R.L., Aber, J.D. & Lajtha, K., 1997: Influence of Excess Nitrogen Deposition on a White Spruce (*Picea* Glauca) Stand in Southern Alaska. Biogeochemistry **38**, 173-187.

Wieler, A., 1905: *Untersuchungen ueber die Einwirkung schwefliger Saeure auf die Pflanzen*, Berlin: Borntraeger.

Xu, Y.-J., Blanck, K., Bredemeier, M. & Lamersdorf, N.P., 1998: Hydrochemical input-output budgets for a clean rain and drought experiment at Solling. *Forest Ecology and Management*, **101**, 295-306.

Section 2

Engineering and planning

APPLICATION OF ARTIFICIAL INTELLIGENCE TO RISK ANALYSIS FOR FORESTED ECOSYSTEMS

Daniel L. Schmoldt[1]

ABSTRACT

Forest ecosystems are subject to a variety of natural and anthropogenic disturbances that extract a penalty from human population values. Such value losses (undesirable effects) combined with their likelihoods of occurrence constitute risk. Assessment or prediction of risk for various events is an important aid to forest management. Artificial intelligence (AI) techniques have been applied to risk analysis owing to their ability to deal with uncertainty, vagueness, incomplete and inexact specifications, intuition, and qualitative information. This paper examines knowledge-based systems, fuzzy logic, artificial neural networks, and Bayesian belief networks and their application to risk analysis in the context of forested ecosystems. Disturbances covered are: fire, insects/diseases, meteorological, and anthropogenic. Insect/disease applications use knowledge-based system methods exclusively, whereas meteorological applications use only artificial neural networks. Bayesian belief network applications are almost nonexistent, even though they possess many theoretical and practical advantages. Embedded systems -that use AI alongside traditional methods- are, not unexpectedly, quite common.

Keywords: *Artificial intelligence, risk analysis, forest ecosystems, hazard, disturbance.*

[1] USDA Forest Service, Southern Research Station,Biological Systems Engineering Dept., 460 Henry Mall, University of Wisconsin, Madison WI 53706-1561
e-mail: dlschmol@facstaff.wisc.edu

K. von Gadow (ed.), Risk Analysis in Forest Management, 49–74.

1 Introduction

Disturbance events are endemic to all complex systems, whether the system is natural or man-made. Complex systems (e.g. ecosystems), that people are dependent on, consist of many, interacting subsystems. When one or more of those smaller subsystems (or their interconnections) fails or behaves aberrantly, it often causes the behavior of the overall ecosystem to change. Such an ecosystem change, then, percolates through to the larger system composed of human populations together with this ecosystem. Many of the institutions (e.g. economic, biological, social, and cultural) inherent to human populations' success and prosperity thrive in a static environment where current conditions are presumed representative of future conditions. Because unpredictable disturbance events modify the trajectory of conditions over time, they can disrupt the status quo of human populations. Therefore, populations must continually exist with uncertainty surrounding future events and their impacts. Life is inherently risky.

Because disturbances cause changes in future scenarios, that differ from status quo expectations, disturbances are often viewed negatively as *loss events*. That is, human populations incur some penalty for disturbances - for being unprepared, "I didn't expect it to rain today, so I didn't take my umbrella. It rained. I got wet. Getting wet greatly reduced my physical comfort level." Disturbances to forest ecosystems are not implicitly "bad" for those ecosystems per se, but they do change their character and the benefits derived from them by human populations. These benefits include commodity values (e.g, timber), amenity values (e.g. scenic vistas), and other ecosystem values (e.g. biodiversity). Even though some disturbances can have beneficial impacts from a human standpoint -for example, wildlife openings created by small, intense fires- the unpredictable nature of disturbances disrupts human expectations of future conditions, and results in a change of benefits and creates losses for those human institutions that expect certain benefits at certain points in time and space. Therefore, for the purposes of risk analysis in this paper, disturbances are cast as "loss events" and viewed in a negative light. As Lackey (1997) noted, risk is almost universally treated as *adverse*, even though ecological change can be either desired or undesired, depending on the value system in place.

Forest ecosystems are subject to a variety of disturbances, including fire, drought, insect and disease attacks, windthrow and breakage, air pollution, rain and surface water acidification, snow/ice damage, threatened and endangered species viability, and other small-scale disturbances. Some of these events are natural and some are anthropogenic. Insect and disease problems have received the greatest attention due to their ubiquitous presence in all types of forests and in all locations; fire is probably the next most important disturbance. Nevertheless, depending on the site, any one of these disturbances can assume the highest priority due to frequency and/or severity. As human populations manage forests and plan for the future, it is important to know how much of a potential problem these disturbances offer, both now and for future generations. That is, "How much are we likely to lose from a potential disturbance?"

The terms "risk" and "hazard" are often used synonymously in the literature (which is consistent with most dictionary definitions), but they are also treated quite vaguely in many cases. Nevertheless, I will try to couch past reports within a more rigorous definition of risk. For this paper, I will adopt the same definition used by others (e.g. Schmucker, 1984; Beer and Ziolkowski, 1995; Lackey, 1999) who treat *risk* as an attribute of a loss event (disturbance), which is composed of 2 components: *potency* (or cost, the severity and extent of the loss event) and *chance* (the likelihood of the loss event). This is also consistent with an actuarial definition that treats risk as expected value loss, where the expected value of a loss event, $E(X_i)$, is the product of the probability of the event, $Pr(X_i)$, and its associated cost $C(X_i)$. For n mutually exclusive events $X_1, X_2, \ldots, X_n$, such that all the $Pr(X_i)$ sum to 1, the total loss is then the sum of the individual expectations $E(X_i)$. When we apply this defintion to past work, however, we will see that different approaches to risk analyze different things. Some examine only potency, and do so in terms of severity only or extent only or both, e.g. the intensity (mortality level) of a pest outbreak versus the geographic spread of the outbreak. This assessment of the undesirable effect is often termed "hazard" (q.v. Beer and Ziolkowski, 1995). In the review that follows, we find that some researchers analyze risk as the *likelihood* of a loss event only, wherein either an event probability is estimated or predisposition for a loss event is assessed. Still other examples provided demonstrate a combined risk assessment that fuses potency and chance, without any explicit actuarial mathematics. While all these different flavors of risk analysis may seem different and lacking commonality, a unifying and underlying objective is to "assess the negative consequences of an uncertain and undesirable event."

Risk analysis can be applied to either current situations or future conditions. Both types of quantitative estimates are difficult to produce, owing to a lack of data, uncertain events, and complex resource interactions. Therefore, artificial intelligence (AI) methods -which are effective in data-poor environments, can handle uncertainty, and use heurisitics to encapsulate complexity (Schmoldt and Rauscher, 1996)- appear well suited to risk analysis problems.

This paper examines several AI techniques for analyzing risk, fully recognizing that the biological science that gives a risk analysis substance and the social context that drives the inquiry into risk analysis (and its interpretation) are outside the scope of this review. AI applications of risk analysis are diverse, due to the breadth of forest-related resources. It would be unrealistic to try to capture them all in a few pages, but this review will rather attempt to identify some key issues that characterize risk in managed ecosystems and how AI methods have attempted to address those issues. The objectives here are two-fold: (1) provide a sythesized overview of the literature in this important area of forest management and (2) describe how the attributes of AI methods make them good candidates for assessing risk. Before delving into the various disturbances addressed by AI risk applications, brief reviews of several important AI methods are provided.

2 Artificial Intelligence Methods

Artificial intelligence methods include a wide variety of computational and algorithmic approaches to problem solving. Their common theme, however, is that they attempt to provide computers with a capability to solve problems in ways that have traditionally been the purview of humans. This means that AI methods incorporate such human cognitive abilities as: reasoning with anecdotal information and best-guess judgment (knowledge-based systems), using vague and commonplace descriptions of objects and events (fuzzy logic), making decisions based on learned patterns with little explicit knowledge or rationale (artificial neural networks), and dealing with large amounts of uncertain, yet interrelated, data and information (Bayesian belief networks). While it is possible to combine these different methods (e.g. fuzzy rule bases and fuzzy neural networks) or to augment with expanded capabilities (e.g. knowledge-based systems that explicitly handle uncertain and inexact information), this overview will treat each method separately and only mention extensions to the standard methods when reviewing particular applications later. Each of the following subsections provides a brief introduction to these techniques in turn.

2.1 Knowledge-Based Systems

A knowledge based system (KBS) is a computer program capable of simulating that element of human knowledge and reasoning that can be formulated into units of knowledge so that a computer can approximate human ability to solve problems (Waterman and Hayes-Roth, 1978; Barr and Feigenbaum, 1982; Schmoldt and Rauscher, 1996). KBSs are distinguished as a subset of AI systems by the fact that they make domain knowledge explicit and separate from the remainder of the system's "reasoning" mechanisms (*reasoning engine*). This separation of knowledge and algorithms is depicted in *Figure 1*, where the *knowledge base* contains domain-specific knowledge and the reasoning engine operates on that data structure to make inferences and draw conclusions. The *working memory* contains data and information pertinent to a specific problem at hand. The reasoning engine then applies the knowledge base to the contents of working memory for problem solving.

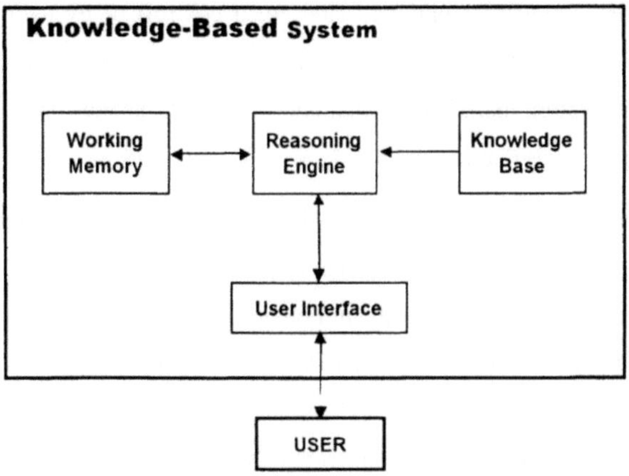

Figure 1. A typical knowledge-based system "reasons" about a particular problem described in its working memory by applying the expertise resident in its knowledge base.

There are numerous techniques for knowledge representation, but traditionally the most common one is the use of *condition-action rules*-see Luger and Stubblefield (1989) and Schmoldt and Rauscher (1996) for a comprehensive review of knowledge representation techniques. Condition-action rules are IF-THEN statements where the consequent action(s) are performed if the premise condition(s) are true, for example, "IF site = sandy AND tree_height < 20 AND previous_winter = mild THEN insect_hazard = high." This method of knowledge representation is popular because each rule is modular and contains a "chunk" of domain knowledge. KBS programmers find rules easy to program, and experts are often able to express their heuristic knowledge in an IF-THEN format.

Working memory is like the short-term memory of a KBS. It contains assertions about the problem currently being investigated. These assertions may be obtained from the user (via queries), from external programs, from a real-time process, from external data files, or inferred from other facts already known. Usually a *closed world assumption* is involved, i.e. only those assertions that are present in working memory are true, all other possible assertions about the state of the world are assumed false.

The knowledge base and working memory are passive entities, whereas the *reasoning engine* navigates through the knowledge base and registers established assertions in working memory. Navigation is performed by the particular control strategy the reasoning engine employs. A control strategy determines the order in which knowledge base elements (e.g. rules) are examined to reach a problem solution. In a rule-based knowledge representation, the inferencing method is usually *modus ponens* and rules are selected for evaluation either by the content of their premise conditions (data-driven control) or by their consequent actions (goal-driven control). Details of how the reasoning engine operates are determined by the knowledge representation method use, what types of assertions must be made, and the overall problem-solving methods that are applied.

Expert systems are distinguished from KBSs in that the knowledge base of an expert system is not derived from generally available (public) knowledge (e.g. textbooks, etc.), but comes from expert specialists in a problem domain and their "private" knowledge of a field. The distinction between KBSs and expert systems is fuzzy at best because the concept of expertise is, itself, not well defined. Consequently, I will generally use the term KBS inclusively, and not worry about the distinction between KBSs and expert systems.

2.2 Fuzzy Sets and Fuzzy Logic

There are several ways in which one can view the nebulous concept of "uncertainty." Whereas statistical probability theory treats uncertainty as randomness or likelihood, it may be more appropriate to consider the *meaning* of information instead. Many of our descriptions of what we know and how we use our knowledge can only be specified in *vague* terms. For example, "short person" and "tall person" are vague descriptors because they are non-explicit and subjective. Zadeh (1978) developed *possibility theory* as a way to treat vague, yet meaningful, information and as an alternative to probability theory's treatment of random events. Possibility theory is an extension of probability theory wherein the "excluded middle" principle is relaxed. That is, an event can belong partially to both the set "true" and the set "false."

Fuzzy sets were introduced by Zadeh (1965) to represent possibility measures. Formally, let S be any set (e.g. the possible heights of trees, in meters), s an element of S, and F a *fuzzy subset* of S (e.g. heights of tall trees). Then the possibility that s is in F (i.e. that a particular height, say 30m, is in the set of heights that are interpreted as "tall" for a particular tree species) is defined by a membership function q. The function q maps all heights onto the interval [0,1]. "Tall" is then referred to as a *fuzzy number* representing the fuzzy set F. A tabulation of $q(s)$ for some values of s in S is presented in *Table 1*.

In keeping with the idea of an algebraic number, we can add, subtract, multiply, and divide fuzzy numbers that are defined over an arbitrary integer base. Such mathematical manipulations of fuzzy numbers were used by Schmoldt (1991) for a qualitative model of soil-plant-atmosphere interactions using the fuzzy arithmetic described by Schmucker (1984) for risk analysis. More often, however, fuzzy sets are used to create fuzzy rule bases.

height *s* in *S* (meters)	**membership *q(s)* in "tall"**
15	0.1
17.5	0.2
20	0.4
22.5	0.5
25	0.7
27.5	0.8
30	0.9
32.5	0.9

Table 1. From possibility theory, all possible heights of trees have an associated possibility, defined by q, that indicates each height's membership in the fuzzy set "tall". All heights less than those listed are assumed to have a membership value of 0 and, similarly, all heights greater than those listed are assumed to have a membership value of 1.

The calculation of belief in a hypothesis H based on a rule of the form A: IF *E* THEN *H* proceed as follows. First, any premise *P* in *E*, will be of the form *X* is *F*, which means that the variable *X* is in the fuzzy set *F*, for example, *X* might be the average height of the dominant trees in a stand and *F* the fuzzy set *tall*. The interpretation of that clause would be that "average tree height is tall" and the membership value *q* associated with the average height, i.e. *X*, and the set *tall* would represent the belief that the average tree height is tall, i.e. $q(X)$. Second, two clauses joined by *and* have a combined belief that is the minimum belief held by the component clauses. In the case of a disjunction of clauses (joined by *or*), the combined belief is the maximum of the component clauses. Third, when *not* is applied to a clause, the resulting belief is 1 minus the membership value of the clause. In general, if *F* is the fuzzy set *tall* and *F'* is the fuzzy set *not tall*, then *not tall* has the membership function $q(F') = 1 - q(F)$. Belief in the final hypothesis *H*, then, is the combined belief of all the evidence *E*. In some cases, however, the rule A may have a belief value attached to its inference (either a numerical value in the range [0,1] or a fuzzy number) that is used to modify the belief contribution of the evidence. Then, the belief in the hypothesis *H*, $B(H)$, equals $B(E) \times B(A)$.

There are some advantages to the fuzzy set approach. The ability to interpret and utilize natural language terms such as "many", "large", "tall", or "more or less" seems consistent with the way that people communicate information and knowledge to each other. Also, fuzzy sets are an example of what is referred to as a *multi-valued logic*, that is, a single term can have several values simultaneously, each one with its own belief level. For example, in addition to the fuzzy set "tall", we may also have a fuzzy set "above average". The membership of 20m in the fuzzy set "above average" might be 0.8, while it is 0.4 in the fuzzy set "tall." So, the height, 20m, could be interpreted as both "above average" and somewhat "tall". By capturing vagueness and permitting multiple-valued entities fuzzy sets mirror some of the inherently inexplicit ways we reason about our world.

2.3 Artificial Neural Networks

Despite substantial success, KBS models of intelligent behavior have been criticized on theoretical as well as practical grounds (Allman, 1989). The KBS paradigm represents a top-down approach to reasoning. This perspective assumes that reality can be described by abstract symbols that represent objects and their relationships to one another. Thought is based on symbols and, indeed, human consciousness is almost synonymous with language, the mental manipulation of symbols (Smith, 1985). While high-level, abstract thinking can often be reasonably formulated in this way (e.g. expert system knowledge), this symbolic model begins to break down when applied to sensory level activities, such as speech recognition, vision, and pattern recognition. KBSs also fail when: problems are highly quantitative, have no well-reasoned or explicable model relating input information to output decisions, or no human specialists exist.

Many of the models and techniques developed in AI borrow from naturally occurring models of complex system organization. The human brain is often used as a model, but cultural/social systems and biological systems (e.g. an ant colony) have provided valuable insights for AI also. Artificial neural networks (ANNs), also referred to as parallel distributed processing by Rumelhart and McClelland (1986), follow this natural/artificial association quite closely by modeling brain nerve cells and their interconnections. The seminal work of developing mathematical models to simulate neural networks was conducted by Hebb (1949).

In response to the limitations of symbolic processing, some AI researchers have turned to recent advances in neurophysiology in their search for more useful models of some aspects of human thought processing. These models signify a bottom-up description of thinking, rather than top down. An artificial neuron is depicted in *Figure 2*. In general, a neuron environment consists of an input vector $\mathbf{X} = x_1, x_2, \ldots$ a weight vector $\mathbf{W} = w_1, w_2, \ldots$, a summation block that combines the inputs and weights, and an activation function f that transforms the weighted sum into an output signal. The net result of applying the weights **W** to the inputs **X** is the value, $net = \mathbf{XW}'$. An activation function f, usually a nonlinear function or a threshold function, is applied to the value *net* to produce an output value, $OUT = f(net)$. Several of these neurons may be placed within a single layer of a network to create what is often called a perception (*Figure* 3a). When perceptions are organized into multiple-layer networks (***Figure*** 3b) and utilize a nonlinear activation function, they become quite powerful computationally (Wasserman, 1989).

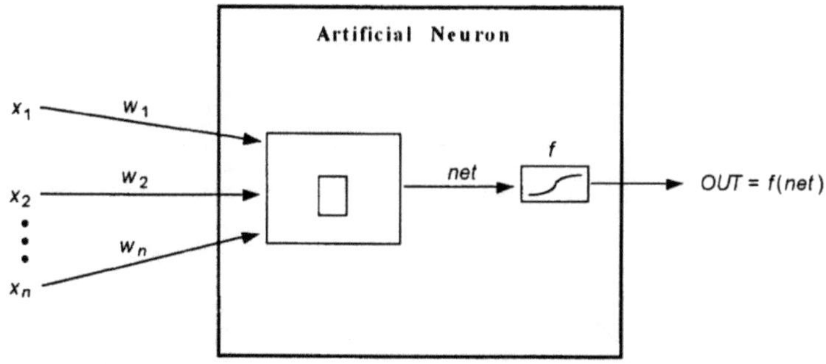

Figure 2. A single artificial neuron contains inputs X, weights W, a summation function, and an activation function f to produce an output response, OUT.

The last illustration (*Figure* 3b) represents the general architecture of a multiple-layer neural network. For a network to operate correctly, it must be trained by adjusting its weight arrays. There are two general types of training methods: supervised and unsupervised. In supervised training, each input vector is paired with a target output vector to produce a training pair. As each input vector is applied to the network, its output is compared to the target and an error is calculated. This error is used to adjust the weights so as to minimize the error. When the errors for the entire training set are acceptably low, then the network has been trained. Unsupervised training requires no target vectors, only input vectors. The training algorithm modifies weight vectors to produce output vectors that are "consistent." Hence, similar input vectors are organized into classes designated by the output patterns. In addition, some transformation may need to be applied to convert the output patterns into intuitively understandable classes. Just as there are many different architectures for neural nets, there are also numerous different training algorithms that can be applied to them. This diversity in architectures and training methods permits a rich and varied repertoire of applications.

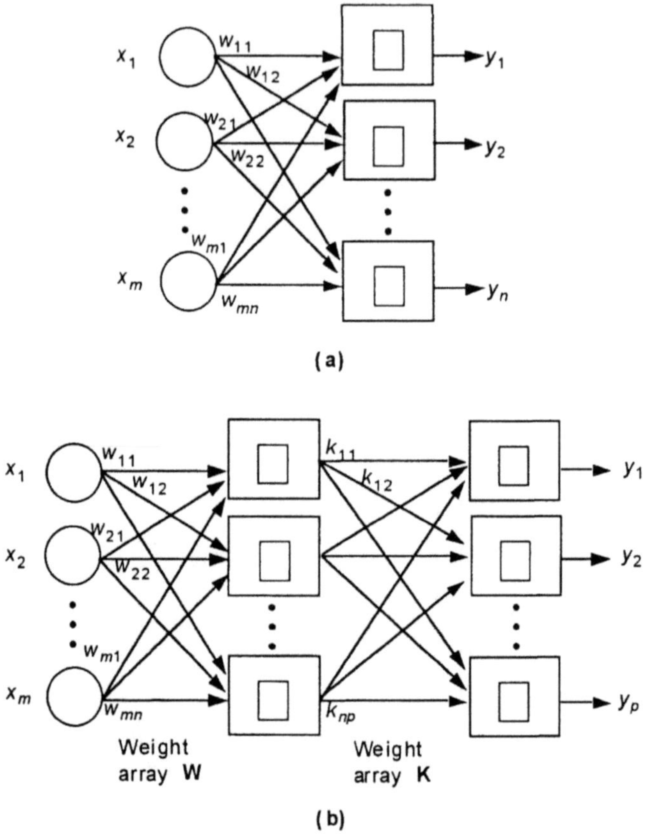

Figure 3. A number of processing units can be arranged into a single layer network (a), termed a perception. More complex and useful networks contain multiple layers (b).

In forestry, empirical statistical models have traditionally been used to relate dependent and independent variables. ANNs can perform the same relationship-finding task, although in a much different way. Despite the past and continued usefulness of traditional statistical models, ANNs have a number of advantages that have proven valuable. In particular, ANNs require no distributional assumptions regarding the data, can discern nonlinear relationships as easily as linear ones, and require no explicit specification of a mathematical model relating input variables to output variables. On the other hand, ANNs possess some significant disadvantages compared to statistical models. First, ANN training uses an interative process and hence requires much longer to reach a solution than a typical maximum likelihood estimation procedure used by linear statistical models. Second, because training is iterative, there is no guarantee that an optimal solution has been reached; often local minima are selected. Third, the final ANN architecture and weight matrices can appear very much like a black box, providing little insight into any true relationship between input and output. Fourth, when ANNs are applied to a specific problem, there are no confidence levels associated with a prediction, as with statistical models. Despite these limitations, ANNs have been found to be superior to logistic regression and discriminant analysis in many instances (e.g. Salchenberger *et al.*, 1992; Tam and Kiang, 1992; Hill *et al.*, 1994; Tomlins and Gray, 1994; Ruisanchez *et al.*, 1996; Marzban *et al.*, 1997). They ususally outperform these classification models (Bansal *et al.*, 1993) when one or more of the aforementioned situations hold, i.e., unknown or non-normal data distributions exist, relationships may be complex and involve many independent variables, linearity of the relationships is unknown (Bansal *et al.*, 1993; Hill *et al.*, 1994), or theoretical/functional models are not apparent (Hill *et al.*, 1994). The reivew of ANN applications in business and finance by Hill et al. (1994), demonstrates broad superiority of ANNs over regression, logistic regression, discriminant analysis, and nonlinear regression. ANNs are not universally better than statistical models, but do possess distinct advantages in some circumstances.

2.4 Bayesian Belief Networks

Much of our understanding of how the world works uses cause-effect models (or influence models). That is, the behavior of any phenomena is understood by considering the influences that other phenomena/events have upon it. For example, we might model tree growth as influenced (caused) by available light, soil moisture, and soil nutrients. The rules that are often used in KBSs -if A and B, then C- are one way of representing and reasoning about these cause-effect relationships.

While measures of uncertainty, e.g. certainty factors (Shortliffe and Buchanan, 1984), can be used within a rule-based structure (e.g. Schmoldt, 1987), interpretation of the rules and interpretation and propagation of the uncertainty measures may not always coincide very well. Furthermore, it is only possible to reason in one direction using rules,whereas it is often useful to hypothesize a *cause* given some knowledge about the presence of an *effect* (reasoning in an effect → cause direction). Finally, if information about A is missing, then the rule "if A and B then C" cannot be used. Bayesian belief networks (BBNs) overcome these limitations of the rule-based formalism by encoding probabilistic relationships among variables in a graphical model (Heckerman and Shachter, 1995). BBNs began with Pearl (1988), who first applied Bayesian probability theory to an influence graph representation of knowledge to allow bi-directional reasoning with uncertainty and subjective knowledge.

The use of Bayes' theorem requires an a priori belief (also called a *prior probability*), $P(H_i)$, in each hypothesis and it also requires conditional probabilities, $P(E|H_i)$, of the current evidence, E, given each possible true hypothesis. This relationship between evidence and hypotheses can be seen in *Figure 4*. Here, we have three mutually exclusive hypotheses H_1, H_2, H_3. Given some evidence, E, each of these hypotheses has some probability of being true. Ultimately, the value we seek is called a *posterior probability* (it results from an updating of the prior with the conditional information) and is represented by the expression, $P(H_i|E)$. In an actual KBSs, the posterior probability is usually an inference of some sort relating the evidence to a particular hypothesis, e.g. an if-then rule. Mathematically, this can be calculated as:

$$P(H_i \mid E) = \frac{P(E \mid H_i) \bullet P(H_i)}{\sum_{k=1}^{n} P(E \mid H_k) \bullet P(H_k)} \tag{1}$$

where the denominator is precisely $P(E)$. Using *Figure 4*, $P(H_i)$ is the percentage of the total area of the big rectangle occupied by the rectangle H_i, and $P(E|H_i)$ is the percentage of the area in H_I that is shaded.

The variables of interest in a BBN are represented in a unidirectional, acyclic graph. This means that causation only goes one way and there are no feedback loops. Under these conditions, it is possible to arrange the network of causation so that any variable in the BBN is conditionally independent with a set of other variables. Independence among variables is essential for the application of Bayes Theorem.

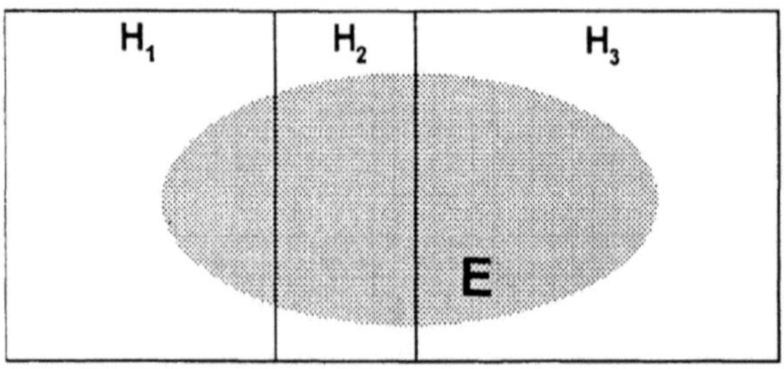

Figure 4. Given three mutually exclusive hypotheses H_1, H_2, H_3, each has some probability of being true a priori, i.e. before E has been observed. The a priori probability of each hypothesis is reflected by its relative area within the total area of the rectangle. Each of these hypotheses has a different probability of being true under the condition of the evidence, E, being true. Therefore, with the observance of evidence, E, the focus is no longer the relative area of each hypothesis within the large rectangle, but rather the relative area of each hypothesis within the shaded oval.

While the example in *Figure 5* is rather trivial it illustrates how one might construct a cause-effect network. The variables *drought*, *tree size*, and *sandy soil*, each have prior probability distributions, whereas *insects present* and *infestation* have conditional distributions. Based on values entered for the former variables, probabilites will propagate and update belief in the latter variables. That is, the probability that an infestation is present can be projected given the other evidence P(*infestation* | *drought* & *tree size* & *sandy soil*). However, it might also be desirable to know P(*drought* | *infestation*), i.e., how likely it is that drought conditions are present given that an infestation has occurred. In that case, this network might be part of a larger network that links drought to fire danger risk, wherein an insect infestation might signal conditions conducive to an escaped fire.

BBNs enable one to utilize both subjective judgement (conditional probabilities) and existing data sources (prior probabilities). Their strong theoretical underpinnings make them attractive to management scientists and operations researchers familiar with decision-analytic techniques.

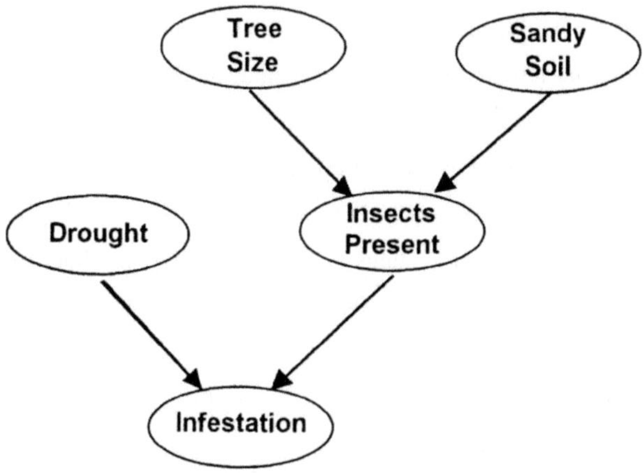

Figure 5. An influence graph illustrates the factors that might cause an insect infestation. Variables that are not directly or indirectly connected by links are conditionally independent.

3 Risk Analysis

As noted above, the literature on risk analysis in natural resources uses risk terminology quite loosely and ambiguously. Risk analysis in one instance may be termed hazard analysis in another, or may address only the *chance* component of risk without any consideration of *potency*. Consequently, it would be difficult to categorize risk applications solely using a particular analysis type. Rather, the following sections examine AI and risk in managed ecosystems according to disturbance type, including fire, insects/diseases, meteorological, and anthropogenic. There are other applications of AI and risk analysis for other disturbances, but these four represent the most common or the most important ones. A couple of instructive examples will be briefly described in each case, with an intent to illustrate the unique contributions that AI has made to risk analysis in each instance.

3.1 Fire

In many parts of the world, fire is the major natural disturbance affecting ecosystems. Fire can produce short-and long-term impacts depending on frequency, intensity, and extent. It impacts not only forest resources, but often human populations directly when urban settlements extend into forested areas. Fire requires two elements: combustible fuels and an ignition source. Risk in this context considers either fuel loading/moisture or ignition likelihood or both factors. Fuel loadings and moisture effect are both risk potency and risk chance. Consequently, there is little commonality between risk terms and fire terms such as behavior, severity, occurrence, loading, moisture content, etc., which are the foci of fire risk analysis. Nevertheless, I will apply these risk terms to fire, adding clarifications as necessary. *Table 2* contains a summary listing of fire risk analysis applications.

The KBS reported by Kourtz (1987) for dispatching fire control resources in Quebec, Canada was one the first applications of AI in fire management. One component of this decision support system evaluates the post-detection fire behavior state and damage potential. This is accomplished using a variety of rule bases and existing simulation models of fire spread. The fire's potency is categorized into one of six behavior situations that partially determines subsequent dispatch of control resources. All the rule bases were developed from the expertise of highly skilled dispatchers; continual modification and evaluation over several fire seasons resulted in a dispatch system that rivaled these human specialists. The final system enables less-experienced dispatchers to apply the knowledge of the best fire dispatchers, along with useful data bases and models, to react intelligently in stressful and time-critical situations.

Application	Analysis Type	AI Method(s)	System
Fire			
Fire situation classification (Kourtz, 1987)	Potency	KBS	Embedded
Wildfire occurence prediction (Vega-Garcia *et al.*, 1996)	Chance	ANN	Stand-alone
Fuel moisture prediction (Ball, 1997)	Potency	ANN	Stand-alone
Prescribed fire escape (Stock, 1996)	Chance	KBS	Stand-alone
Broad-scale fire severity (Lenihan *et al.*, 1998)	Potency	KBS	Embedded

Table 2. Applications of AI to risk analysis for fire disturbances are summarized across several attributes, including type of risk analysis, AI method(s) used, and stand-alone/embedded system.

While some risk analyses treat risk factors using some extant theory of their importance and impact, other analyses attempt to generalize past data without regard to any underlying risk model. An example of this approach is daily wildfire occurrence prediction (Vega-Garcia *et al.*, 1996) using historical data for specific zones within a provincial forest in Alberta, Canada. Rather than assigning likelihood to an event (chance), this study predicts actual event occurrence, but we will still refer to the prediction as *chance* (*Table 2*). A fire-weather index, zone area, and zone identifier were used as inputs to an ANN that was trained on data from 5 fire seasons and predicted *fire* or *no-fire*. Testing of the ANN on data from 2 other fire seasons gave 85% prediction accuracy for no-fire observations and 78% accuracy for fire observations. This type of model implicitly includes both fuel and ingnition information, as they interact to determine fire occurrence.

In another application of ANNs, Ball (1997) attempted to predict fuel moisture from various environmenal variables in the desert areas of the U.S. southwest. Separate ANNs were created to predict live- and dead-fuel moisture. Because this context is prescribed burning and fire behavior, fuel moisture prediction can be considered risk potency. Only 51 training samples were used, and when both ANNs were tested with these same data, they predicted moisture accurately in 92% of the cases. When splitting the same data set into training *and* testing sets, much lower accuracy was obtained. This suggests that their 92% value is an overestimate of the true accuracy. Given the small sample size, cross-validation techniques should have been used (Weiss and Kulikowski, 1991). No rigorous testing was performed with separate data sets, although a visual comparison was made between simulated fuel-moisture maps and ANN-predicted fuel moisture maps. Again the results were not impressive. The authors suggested that the paucity of sample data and the use of a single fuel loading value for all fuel sizes greatly limited their results. Quantitative AI methods, such as ANNs, are not unlike other quantitative methods in that the method cannot compensate for poor data.

A related treatment of prescribed fire by Stock et al. (1996) analyzed the chance of escape of a prescribed burn. Their likelihood estimate uses a scale of 0 (no chance) to 100 (extreme escape danger), and is separate from any consideration of escape consequences (potency). Their KBS attempted to improve on a matrix model, previously developed and currently used by fire managers, by incorporating less explicit and more nonlinear aspects of the prescribed burning process. They used 3 test case comparisons with expert predictions to identify areas where their KBS could be modified/improved. In one of those cases, they found that the human expert was not separating chance (escape likelihood) and potency (escape consequences), as their KBS was attempting to do. By enabling fire managers to make this distinction in actual operations, they hope to place prescribed burning on a more objective decision-theoretic foundation, leading to better decisions.

A slightly different use of AI in risk analysis is the Dynamic Global Vegetation Model (DGVM, Lenihan *et al.*, 1998). This model simulates global change impacts from extreme fire events. The AI component does not directly predict either fire chance or potency, but rather generates a spatio-temporal distribution of vegetation classes using a rule base. These vegetation types are inputs to the fire simulation component of DGVM that projects fire behavior, severity, and effects. DGVM allows researchers and policy makers to predict changes in the occurrence and impacts of extreme fire events under different global climate change scenarios.

3.2 Insects and Diseases

The only natural disturbances more ubiqitous than fire are those that result from pathogens and insects. These disturbances occur in every forest type around the globe. Managed forests are particularly susceptible to their impacts due to unnatural management activities, limited host species genotypes, and uniform forest vegetation structure. Both insects and diseases are continually present in forests at sub-epidemic levels. When conditions are rife, however, mere presence can change to epidemic with dramatic, and often catastrophic, consequences. Risk analysis in this context assesses or predicts epidemic-favorable conditions. *Table 3* lists the AI applications covered below.

The first reported AI application in this area is the hazard assessment KBS of Schmoldt (1987). Here, stand hazard rating was an ancillary component of the PREDICT diagnosis KBS system that contained insect and disease knowledge for red pine (*Pinus resinosa*) in Wisconsin (USA). Ratings of low, moderate, or high hazard (along with a numerical estimate of hazard assessment reliability) are provided based on forest location and soil/site factors. Given forest conditions, this assessment provides a qualitative rating of the likelihood that an outbreak will occur.

Rust (1988) used a KBS to rate white pine (*Pinus monticoa*) planting sites in the Pacific Northwest (US) according to their likelihood to develop white pine blister rust (*Cronartium ribicola*). In predicting future blister rust hazard, the system uses site factors and alternative host suitability. It also makes planting recommendations following hazard rating.

Application	Analysis Type	AI Method(s)	System
Insects & Diseases			
Hazard assessment for red pine (Schmoldt, 1987)	Chance	KBS	Stand-alone
White pine blister rust hazard rating (Rust, 1988)	Chance	KBS	Stand-alone
Hemlock looper DSS (Power and Saarenmaa, 1995)	Chance & Potency	KBS	Embedded
Jack pine budworm DSS (Loh *et al.*, 1991)	Chance & Potency	KBS	Embedded
Mountain pine beetle population prediction (Downing and Bartos, 1991)	Chance	KBS	Stand-alone
Spruce beetle risk and hazard (Reynolds and Holsten, 1994; Reynolds and Holsten, 1996)	Chance & Potency	KBS	Embedded
Gypsy moth risk assessment (Potter *et al.*, 2000)	Chance & Potency	KBS	Stand-alone

Table 3. Applications of AI to risk analysis for insect and disease disturbances are summarized across several attributes, including type of risk analysis, AI method(s) used, and stand-alone/embedded system.

The decision support system (DSS) described by Power and Saarenmaa (1995) is unique in this overview in that the authors converted an existing DSS for hemlock looper (*Lambdina fiscellaria fiscellaria*) outbreak and impact prediction in Newfoundland (Canada). They used object-oriented techniques to create a more generic DSS framework that could be used/augmented for other forest pests. Their approach treats the decision-support problem in a more structured manner, while still making use of existing simulation models, data bases, and geographic information system (GIS) layers. In doing so, the authors highlight many of the benefits of using AI approaches for risk analysis, such as object-oriented programming (originally developed in the AI research community).

The jack pine budworm (*Choristoneura pinus*) DSS reported by Loh et al. (1991) is similar in intent and function to the preceding system. Object-oriented techniques are used to link rule bases, a data base, and a GIS to manage jack pine (*Pinus banksiana*) forests in the Lake States (US) and Canada. One of the rule bases generates hazard ratings (low, moderate, high), that seem to incorporate both outbreak chance and impact (potency). Ratings are subsequently used to recommend management options.

Because pests exist continually in forests at endemic levels, it is important to predict when those levels might explode to epidemic levels. Downing and Bartos (1991) developed a KBS to predict mountain pine beetle (*Dendroctonus ponderosae*) population increases in lodgepole pine (*Pinus contorta*) stands in the western US. Climatic and site factors were used to predict the next year's brood level, along with a certainty-factor estimate of confidence in that prediction. Interestingly, the authors developed and used a knowledge-acquisition program to elicit knowledge from experts and then used commercial rule induction software to generate the rules. Separate rule bases were developed by each expert, and the experts then criticized the rule bases of their colleagues. This provides an interesting approach to incorporating multiple experts' knowledge into a KBS.

SBexpert helps manage spruce beetle (*Dendroctonus rufipennis*) outbreaks in south-central Alaska (US) spruce stands (*Picea* spp.). It contains rule bases to evaluate both chance (Reynolds and Holsten, 1994) and potency (Reynolds and Holsten, 1996) of spruce beetle outbreaks. Potency is treated as severity and is measured as percent basal area loss of spruce. SBexpert, version 2, is currently available (http://www.fsl.orst.edu/usfs/sbexpert/), and version 3 will include automated landscape-scale analysis using a GIS.

A web-based KBS (GyMEs) has been developed by Potter et al. (2000) to assess risk for gypsy moth (*Lymantria dispar*) outbreaks in the eastern US. While not the first web-based KBS, it is the first web-based risk analysis application in forestry. GyMEs assesses both potency of an outbreak and likelihood of an outbreak incident to arrive at a risk rating. Site factors, stand condition, disturbance history, and stand susceptibility are used to arrive at a qualitative estimate of risk, e.g. low, moderate, high.

3.3 Meteorological Phenomena

Weather events can cause regular and significant disturbances in forest ecosystems. Severe storms accompanied by strong winds, heavy rains, and ice formation can create major damage to trees and other forest resources. Weather patterns also interact with other disturbances to modify their occurrence or severity, such as fire (lightning strikes, drought) and air pollution (temperature inversions). In some regions of the world, weather events can be a major contributor to losses in the forest, and their spatio-temporal unpredictability makes them particularly risky. While there exists considerable research involving AI and meteorology (q.v. Christopherson, 1997; Gardner and Dorling, 1998), the number of studies with importance to risk in managing forests is much less. *Table 4* summarizes several example AI applications.

Lakshmanan and Witt (1997) developed a fuzzy logic system to predict a critical precursor for severe thunderstorms. Their system automatically analyzes radar imagery and estimates fuzzy numbers from the data, which are then used in the fuzzy rules of a classifier. This classifier labels the critical regions in a radar image as either thunderstorm precursors, marginal precursors, or non-precursors. Their measure of success compares very favorably with other meteorological approaches to identifying thunderstorm precursors.

Application	Analysis Type	AI Method(s)	System
<u>**Meteorological Phenomena**</u>			
Detecting sever e updraft conditions (Lakshmanan and Witt, 1997)	Chance	Fuzzy logic	Stand-alone
Damaging wind prediction (Marzban and Stumpf, 1998)	Chance	ANN	Stand-alone
Precipitation forecasting (Hall *et al.*, 1999)	Chance & Potency	ANN	Stand-alone
Localized precipitation forecasting (Kuligowski and Barros, 1998)	Potency	ANN	Stand-alone
Lightning strike prediction (Frankel *et al.*, 1995)	Chance	ANN	Stand-alone

Table 4. Applications of AI to risk analysis for meteorological events are summarized across several attributes, including type of risk analysis, AI method(s) used, and stand-alone/embedded system.

Certain regions of the world are prone to high-wind damage, which can have dramatic effects even in forested areas. Following prior work that compared ANN performance with that of discriminant analysis and logistic regression (Marzban and Stumpf, 1996; Marzban *et al.*, 1997), Marzban and Stumpf (1998) developed an ANN to predict the occurrence of damaging winds (> 25m/s). Input data were taken from circulation patterns detected by the National Weather Service (US); the ANN model outputs a binary decision. Using a variety of performance measures, the authors found the ANN to be a reliable and discriminating predictor.

Accurate precipitation forecasts can identify the potential for heavy rains and associate flash flooding that can create extreme hydrologic events in watersheds and can damage man-made structures. Hall et al. (1999) developed 2 separate ANNs for rainfall prediction, one for precipitation occurrence probability and one for rainfall amount. Predictons were made for mean daily rainfall in a 5000 km^2 metropolitan area (presumably the same approach would work at the watershed scale). Both ANNs were extremely accurate in validation tests that covered daily precipitation amounts for a two-year period.

Because current quantitative precipitation forecasting models operate at a large scale, they are unable to make accurate localized spatio-temporal forecasts. Regression models have traditionally been used to interpolate for local regions and shorter time frames. Kuligowski and Barros (1998) developed an ANN to substitute for existing linear regression models. Tests conducted at four locations in the middle Atlantic region of the US showed that their ANN's accuracy performed comparably with forecasts generated using regression methods, especially for important heavy rainfall events.

Accurate spatio-temporal lightning strike prediction can aid forest managers to prepare for forest fires. Frankel et al. (1995) describe the development of an ANN to forecast lightning around the Kenndey Space Center in South Florida (US). Input data included a variety of detailed meteorological variables for the immediate area. Prediction results for 0-15, 15-30, 30-60, and 60-120 minute time windows within subareas of the Center were numerically equal to total-area predictions using a traditional linear-regression prediction model. Although the input data for their ANN were much more detailed than could easily be obtained in forested areas, the ANN's ability to forecast lightning events over small spatio-temporal scales holds promise for similar efforts in forests using readily available meteorological data.

3.4 Anthropogenic Effects

An ever burgeoning human population and its migration into, and adjacent to, previously sparsely inhabited forested areas is a stressor for forest resources. While populations can disturb ecosystems at a distance through atmospheric pollutants, water pollution and misuse, grazing, greenhouse-gas induced climate change, and wildlife habitat disruptions, they can also impact ecosystems proximally by their land management and recreational activities. Such anthropogenic influences modify the course of ecosystem development and generate substantial risk, just as fire and insects. *Table 5* lists several AI applications for anthropogenic risk.

Application	Analysis Type	AI Method(s)	System
Anthropogenic Effects			
Lake acidification assessment (Lam *et al.*, 1989)	Potency	KBS	Embedded
Pesticide risk (Messing *et al.*, 1989)	Potency	KBS	Embedded
Recreation use prediction (Pattie, 1992)	Potency	ANN	Stand-alone
Air pollution forecasting (Nunnari *et al.*, 1998)	Potency	ANN	Stand-alone
Water quality monitoring (Varis *et al.*, 1993)	Potency	BBN	Embedded

Table 5. Applications of AI to risk analysis for anthropogenic disturbances are summarized across several attributes, including type of risk analysis, AI method(s) used, and stand-alone/embedded system.

An interesting approach to assessing the extent of lake acidification is provided by Lam et al. (1989). The authors embedded a small KBS within a simulation environment (RAISON) to automatically select the appropriate acidification model and to analyze model results for anomalous behavior. In a comparison using 34 lakes in Quebec, Canada, the knowledge-based combined model of RAISON performed comparably to the individual component models. Proper application of simulation models is difficult, even for experienced users. A KBS that can elevate human performance in this task is a significant improvement over manual procedures.

While the KBS by Messing et al. (1989) deals with an agricultural application, it is worth mentioning briefly because it addresses pesticide risk - which is an issue in forestry also. Herbicide and pesticide applications to forests are relatively common in industrial operations, but little attention is often given to the potential damage those chemicals might cause for a pest's natural enemies, soil organisms, etc. The authors' rule base links to data bases, simulation models, and graphical tools to help the user estimate mortality for beneficial organisms.

Recreation over-use can have significant impacts, especially in wilderness areas. The ability to forecast visitor usage can enable managers to establish protocols and policies to protect sensitive resources and to preserve the character of the wilderness. Pattie (1992) proposed an ANN to predict overnight backcountry stays in national parks and recreational visitor-days in national forests (US). Input variables selected for use in the ANN were intended to reflect the factors that drive recreation use/non-use. No information is available on the results of this development effort, but the proposed approach warrants inclusion here due to the anthropogenic risk (recreation) it deals with and to the AI technique selected.

Many forested regions are downwind from significant sources of atmospheric pollutants. These pollutants are known to have dramatic impacts on vegetation (ozone exposure) and water resources (nitrogen and sulphur deposition). Nunnari et al. (1998) used time-series data and an ANNs to predict short- and medium-range concentrations of ozone and nitrogen oxides. Their ANN provided improved forecasting over traditional statistical models (auto-regressive moving average with exogenous inputs), in particular for predicting ozone values exceeding a threshold.

Water resources health is an important aspect of forest ecosystems. Issues of monitoring system functioning and cost, in addition to risk attitudes and discount rates, were included in a BBN model of water quality monitoring (Varis *et al.*, 1993). The authors' model analyzed the costs and benefits of different monitoring systems under various scenarios of risk aversion, discount rate, and cost uncertainty. The flexibility of this type of model allows decision makers to form choices based on preferences (risk attitude) and exogenous variables (discount rate), in addition to scientific information (monitoring equipment).

4 Conclusions

While this overview of AI applications for risk analysis of forest resources does not review the literature comprehensively, it is, nonetheless, reasonably representative of the things that have been done. Consequently, there are some concluding observations that can be made regarding this area of research and application.

The integration of KBSs with simulation models is a natural, and not unexpected, approach toward risk analysis. Risk assessment often requires the use of quantitative values and many simulation models already exist for generating these values. KBSs, then, allow one to use that data more effectively by combining it with less quantitative information or to select the best modelling approach when several are available or to interpret model output.

There are several interesting observations apparent in the application summaries (Figures 2-5). Regarding analysis type, fire applications are split between chance and potency, insect/disease applications always include chance (often combined with potency), meteorological applications are mostly concerned with event occurrence chance, and anthropogenic disturbance applications consider potency only. These particular emphases reflect obvious disturbance differences, e.g. anthropogenic disturbances exist, so the question of chance is moot and one wishes to predict severity and extent.

Regarding AI methods used, fire applications use both KBSs and ANNs, insect/disease applications use KBSs exclusively, meteorological applications use ANNs almost exclusively, and anthropogenic applications use a mixture of all four AI methods. The single-minded use of KBSs for insect and disease disturbances may be due to a lack of quantitative data and relationships regarding pests and forests. Being limited to qualitative information does not preclude the use of ANNs and BBNs, but it may bias researchers toward rules and away from more quantitative methods (Bayesian statistics, for instance). On the other hand, meteorology has historically dealt with large amounts of data, so the relatively exclusive application of ANNs to model quantitative relationships and to replace existing statistical approaches is not unexpected. Regarding system type, ANNs tend to be stand-alone applications. Although, because a trained ANN can be encoded as a single function call in a programming language, I would expect to see ANNs embedded in other software systems more often in the future. KBSs can be either stand-alone or embedded depending on whether software tools (e.g. simulation models, GISs) are available or needed. Hopefully, as future AI systems for risk analysis are created, developers will take some of these issue into account during system design.

In general, the applications of BBNs for risk analysis are few in number. Two likely reasons for this scarceness are immediately apparent: (1) BBNs are technically more challenging to develop than KBSs and (2) user-friendly commercial software has only recently become available to help risk analysts create BBNs (http://bayes.stat.washington.edu/almond/belief.html), whereas comparable ANN and KBS software has been around for many years. The presentation of BBNs in an ecosystem management context by Olson (1990) is a strong argument for more extensive use of this technology. The integration of a probability model (combining both subjective judgement and available data) with a causal network representation in BBNs makes them very appropriate for risk analysis tasks.

Disturbances in forests are, by definition, changes to the status quo. In doing so, they penalize human populations, that rely on constancy for survival. While forest ecosystem disturbances are a certainty, exactly when and where they occur and how severe their effects are uncertain events. This generates risk. Unfortunately, there is much we do not know about what causes disturbances and how they will impact forest resources. In most cases, neither do we have good data to make predictions. The AI techniques covered here offer some useful tools for dealing with uncertain/missing knowledge and data. In many cases, these methods can augment more traditional approaches and, in other cases, can replace previous methods due to their improved predictions, ease of use, or theoretical advantages.

5 Literature Cited

Allman, W. F., 1989: *Apprendices of wonder.* Bantam, New York.

Ball, B. J., 1997: Fuel moisture prediction in homogeneous fuels using GIS and neural networks. *AI Applications* 11(3): 73-78.

Bansal, A., Kauffman, R. J. and Weitz, R. R., 1993: Comparing the modeling performance of regression and neural networks as data quality varies: A business value approach. *Journal of Management Information Systems* 10(1): 11-32.

Barr, A. and Feigenbaum, E. A. (Eds.), 1982: *The handbook of artificial intelligence.* William Kaufmann, Inc., Los Altos CA

Beer, T. and Ziolkowski, F., 1995: Environmental risk assessment: an Australian perspective. Commonwealth of Australia, Supervising Scientist Report 102 http://www.environment.gov.au/ssg/pubs/risk/risk_foreword.html.

Christopherson, D., 1997: Artificial intelligence and weather forecasting: an update. *AI Applications* 11(1): 81-93.

Downing, K. and Bartos, D. L., 1991: AI methods in support of forest science: Modelling endemic level mountain pine beetle population dynamics. *AI Applications* 5(2): 105-115.

Frankel, D. S., Draper, J. S., Peak, J. E. and McLeod, J. C., 1995: Artificial intelligence needs workshop 4-5 November 1993, Boston MA. *Bulletin of the American Meteorological Society* 76(5): 728-738.

Gardner, M. W. and Dorling, S. R., 1998: Artificial neural networks (the multilayer perceptron)- A review of applications in the atmospheric sciences. *Applications in the Atmospheric Sciences* 32(14-15): 2627-2636.

Hall, T., Brooks, H. E. and Doswell III, C. A., 1999: Precipitation forecasting using a neural network. *Weather and Forecasting* 14(3): 338-345.

Hebb, D. O., 1949: *Organization of Behavior.* Science Editions, New York.

Heckerman, D. and Shachter, R., 1995: Decision-theoretic foundations for causal reasoning. *Journal of Artificial Intelligence Research* 3: 405-430.

Hill, T., Marquez, L., O'Connor, M. and Remus, W., 1994: Artificial neural network models for forecasting and decision making. *International Journal of Forecasting* 10: 5-15.

Kourtz, P., 1987: Expert system dispatch of forest fire control resources. *AI Applications* 1(1): 1-8.

Kuligowski, R. J. and Barros, A. P., 1998: Localized precipitation forecasts from a numerical weather prediction model using artificial neural networks. *Weather and Forecasting* 13(4): 1194-1204.

Lackey, R. T., 1997: If ecological risk assessment is the answer, what is the question? *Human and Ecological Risk Assessment* 5(6): 921-928.

Lackey, R. T., 1999: Radically contested assertions in ecosystem management. *Journal of Sustainable Forestry* 9(1/2): 21-34.

Lakshmanan, V. and Witt, A., 1997: A fuzzy logic approach to detecting severe updrafts. *AI Applications* 11(1): 1-12.

Lam, D. C. L., Swayne, D. A., Story, J. and Fraser, A. S., 1989: Watershed acidification models using the knowledge-based systems approach. *Ecological Modelling* 47(1/2): 131-152.

Lenihan, J. M., Daly, C., Bachelet, D. and Neilson, R. P., 1998: Simulating broad-scale fire severity in a dynamic global vegetation model. *Northwest Science* 72: 91-103.

Loh, D. K., Connor, M. D. and Janiga, P., 1991: Jack pine budworm decision support system: A prototype. *AI Applications* 5(4): 29-48.

Luger, G. F. and Stubblefield, W. D., 1989: *Artificial intelligence: structures and strategies for complex problem solving.* Benjamin/Cummings Publishing Co., Redwood City CA.

Marzban, C., Paik, H. and Stumpf, G. J., 1997: Neural networks vs. Gaussian discriminant analysis. *AI Applications* 11(1): 49-58.

Marzban, C. and Stumpf, G. J., 1996: A neural network for tornado prediction based on Doppler radar-derived attributes. *Journal of Applied Meteorology* 35: 617-626.

Marzban, C. and Stumpf, G. J., 1998: A neural network for damaging wind prediction. *Weather and Forecasting* 13(1): 151-163.

Messing, R. H., Croft, B. A. and Currans, K., 1989: Assessing pesticide risk to arthropod natural enemies using expert system technology. *AI Applications* 3(2): 1-12.

Nunnari, G., Nucifora, A. F. M. and Randieri, C., 1998: The application of neural techniques to the modelling of time-series of atmospheric pollution data. *Ecological Modelling* 111: 187-205.

Olson, R. L., Wagner, T. L. and Willers, J. L., 1990: A framework for modeling uncertain reasoning in ecosystem management 2. Bayesian belief networks. *AI Applications* 4(4): 11-24.

Pattie, D. C., 1992: Using neural networks to forecast recreation in wilderness areas. *AI Applications* 6(2): 57-59.

Pearl, J., 1988: *Probabilistic reasoning in intelligent systems: networks of plausible inference*. Morgan Kauffman, San Mateo CA.

Potter, W. D., Deng, X., Li, J., Xu, M., Wei, Y., Lappas, I., Twery, M. J. and Bennett, D. J., 2000: A web-based expert system for gypsy moth risk assessment. *Computers and Electronics in Agriculture* 27(1-3): 95-105.

Power, J. M. and Saarenmaa, H., 1995: Object-oriented modeling and GIS integration in a decision support system for the management of eastern hemlock looper in Newfoundland. *Computers and Electronics in Agriculture* 12(1): 1-18.

Reynolds, K. M. and Holsten, E. H., 1994: Relative importance of risk factors for spruce beetle outbreaks. *Canadian Journal of Forest Resources* 24: 2089-2095.

Reynolds, K. M. and Holsten, E. H., 1996: Classification of spruce beetle hazard in Lutz and Sitka spruce stands on the Kenai Peninsula, Alaska. *Forest Ecology and Management* 84: 251-262.

Ruisanchez, I., Potokar, P. and Zupan, J., 1996: Classification of energy dispersion x-ray spectra of mineralogical samples by artificial neural networks. *Journal of Chemical Information and Computer Sciences* 36(2): 214-220.

Rumelhart, D. E. and McClelland, J. L., 1986: *Parallel Distributed Processing: Explorations in the Microstructure of Cognition*. The MIT Press, Cambridge MA.

Rust, M., 1988: White pine blister rust hazard rating: An expert systems approach. *AI Applications* 2(2-3): 47-50.

Salchenberger, L. M., Cinar, E. M. and Lash, N. A., 1992: Neural networks: A new tool for predicting thrift failures. *Decision Sciences* 23(4): 899-916.

Schmoldt, D. L., 1987: Evaluation of an Expert System Approach to Forest Pest Management of Red Pine (*Pinus resinosa*). Ph.D. dissertation. University Microfilms International.

Schmoldt, D. L., 1991: Simulation of plant physiological processes using fuzzy variables. *AI Applications* 5(4): 3-16.

Schmoldt, D. L. and Rauscher, H. M., 1996: *Building Knowledge-Based Systems for Natural Resource Management*. Chapman & Hall, New York.

Schmucker, K. J., 1984: *Fuzzy sets, natural language computations, and risk analysis*. Computer Science Press, Rockville MD.

Shortliffe, E. H. and Buchanan, B. G., 1984: A model of inexact reasoning in medicine. In B. G. Buchanan and E. H. Shortliffe (eds.), Rule-based expert systems. Reading MA, Addison-Wesley, p. 232-262.

Smith, C. G., 1985: *Ancestral voices: language and the evolution of human consciousness*. Prentice-Hall, Inc., Englewood Cliffs NJ.

Stock, M., 1996: Estimating the risk of escape of prescribed fires: an expert system approach. *AI Applications* 10(2): 63-73.

Tam, K. Y. and Kiang, M. Y., 1992: Managerial applications of neural networks: The case of bank failure predictions. *Management Science* 38(7): 926-947.

Tomlins, K. I. and Gray, C., 1994: Prediction of quality and origin of black tea and pine resin samples from chromatographic and sensory information: evaluation of neural networks. *Food Chemistry* 50: 157-165.

Varis, O., Kløve, B. and Kettunen, J., 1993: Evaluation of a real-time forecasting system for river water quality. *Environmental Monitoring and Assessment* 28(201-213):

Vega-Garcia, C., Lee, B. S., Woodard, P. M. and Titus, S. J., 1996: Applying neural network technology to human-caused wildfire occurrence prediction. *AI Applications* 10(3): 9-18.

Wasserman, P. D., 1989: *Neural computing: theory and practice*. Van Nostrand Reinhold, New York.

Waterman, D. A. and Hayes-Roth, F., 1978: An overview of pattern-directed inference systems. In D. A. Waterman and F. Hayes-Roth (eds.), Pattern-directed inference systems. New York, Academic, p. 251-269.

Weiss, S. M. and Kulikowski, C. A., 1991: *Computer Systems That Learn*. Morgan Kaufmann Publishers, Inc., San Mateo, 223 pp.

Zadeh, L. A., 1965: Fuzzy sets. *Information and Control* 8: 338-353.

Zadeh, L. A., 1978: Fuzzy sets as a basis for a theory of possibility. *Fuzzy Sets and Systems* 1: 3-28.

RISK ANALYSIS AS A TOOL FOR POLICY EVALUATION: A MULTI-LEVEL COMPARISON EXAMPLE

K. Hollenstein[1]

ABSTRACT

Land-use decisions in forestry are increasingly difficult to make, and tools are required to assess the consequences of different options across time and space. Risk analysis, although traditionally used for other purposes, can address these issues in a transparent and quantitative way. The methodology is applied to a fictional forest enterprise in Switzerland, and the analysis is performed for ownership type, the national and the global level. Two scenarios, no wood harvest and sustainable harvest, are investigated both for normal operation and operation after a catastrophic event. For the no harvest scenario, a substitution of the production elsewhere is also investigated. Six risk variables are taken into account: loss of life, disturbance of human health, loss of ecosystem functionality, loss of soil fertility, greenhouse gas emission and financial losses. The results show that with the assumed trade-off relations, greenhouse gas emission becomes a crucial factor. For the normal operation, the total aggregated risk is lowest with the harvest scenario, whereas for the operation after catastrophe, a substitution from within Switzerland or adjacent countries is less risky because of the increased accident frequency when harvesting salvage wood. The assumptions for the raw data and the trade-off between risk variables are critical for the outcome of the analysis, and changing these input data can change the ranking of options. However, the primary benefit of using risk analysis for policy evaluation is the disclosure of risk transfers that are often not accounted for in other analysis techniques.

Keywords: ***Risk analysis, policy evaluation, forest operations, land use***

---

[1] Department of Forest Engineering, Swiss Federal Institute of Technology Zurich, HG G 21.5, ETH Zentrum, CH-8092 Zurich, Switzerland, Tel +41 1 632 3238; Fax +41 1 632 1146

K. von Gadow (ed.), Risk Analysis in Forest Management, 75–112.

Introduction

Increasing resource consumption due to changes in population and lifestyle on the one side and a non-increasing or decreasing landbase for resource production make land-use decisions for forests more and more difficult. Today, the traditional multiple-use concept that was previously accepted in Europe and North America is often questioned or criticized; and there is public pressure to manage at least a part of the forest landbase for specific purposes such as habitat protection (see for example PSW 1996, Ruth et al. 1994). However, in these discussions, it is not always evident that the long-term consequences and external effects of management or policy decisions are properly assessed not only within, but also across the boundary of the forest system under consideration. As a result, forest managers are permanently confronted with the problem of potential land-use conflicts and the probability of suboptimal or unfair policies.

To rationalize the discussion and contribute to the solution of the problem, it is useful to have tools available that provide an unbiased or at least a reproducible evaluation of different forest policy and management options (i.e. a methodology to perform a prescriptive policy analysis for assessing the policy-making process, cf. Weiss 1997). Due to the nature of the consequences of a decision in forestry, it is necessary that the tools are able to assess and compare the options i) over a long period of time and ii) over a wide range of spatial extents. The temporal dimension can be addressed using statistical terms (expectation, frequency, probability), and relative values (per area or per capita) allow to consider the spatial dimension.

Risk analysis is a discipline that incorporates both these principles. The definition of risk as the product of frequency and intensity accounts for the temporal aspect; the spatial part is covered if the risk is treated as either an emission from or an immission onto a certain area (Fig. 1). In other words, risk analysis can not only be used as a method for investigating the safety of technical or environmental systems, but also as a tool for evaluating management alternatives. The risk analysis methodology provides the formal framework for this special application; the negative effects and consequences of an option are defined as risks. From a rational point of view, the best option is then the one that has the smallest associated risk while still fulfilling the management objectives. Such an approach has several advantages:

i) it is applicable to a wide variety of settings without the need to change the methodology;
ii) it provides a high degree of transparency;

iii) the methodology is generally well accepted,
iv) it can easily be automated and thus repeated at low cost to test for sensitivity.

Of course, there are also some disadvantages in using risk analysis for policy evaluation, namely:

i) the need to provide numerical trade-offs between different values;
ii) the questionable acceptance of a single risk value as a measure for a complex set of consequences and effects.

However, any evaluation technique that uses aggregation has the same shortcomings, and based on the advantages mentioned, it seems at least worthwhile to investigate the potential of risk analysis for this purpose. The need to disclose and operationalize preferences to develop trade-offs has an additional benefit. It requires a positive formulation of values. Rather than deciding on the basis of what is to be prevented, policies have to be defined in terms of what is to be achieved. Generally, this will reduce the probability of involuntarily omitting external effects of a selected policy.

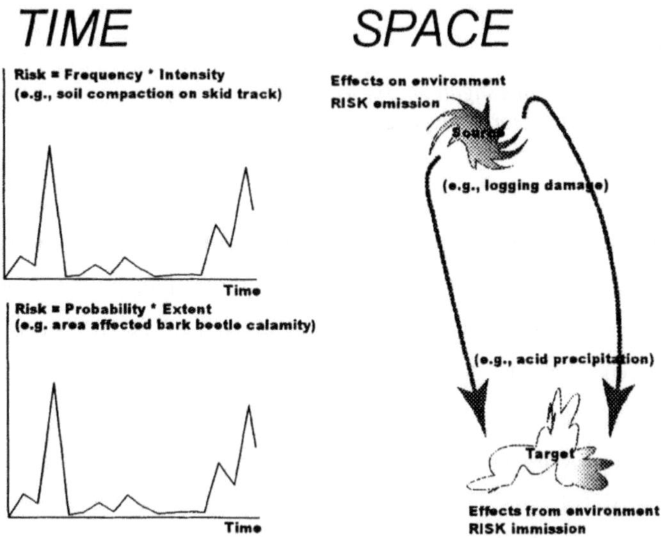

Figure 1. Risk analysis as a methodology covering both temporal and spatial dimensions. The definition of risk as a product of intensity (or magnitude) and probability (or frequency) accounts for the temporal, the concepts of emission and immission for the spatial aspects.

This present article compares two forest management policies for forests in Switzerland with regard to their risk structure as an example application of risk analysis for policy evaluation. The first policy is based on a scenario of sustainable wood harvest that does not noticeably degrade the biomass production potential of a site, but nonetheless changes its vegetative cover. The second policy is an exclusion of any kind of biomass harvest. Both scenarios are evaluated at different decision levels (i.e. forest owner, nationwide, worldwide) and under the assumption of periodic catastrophic events (Fig. 2). The scenarios on the ownership level are applied to a fictive forest in Switzerland, but based on actual forestry statistics (for data such as economic performance, accident frequency, environmental impacts).

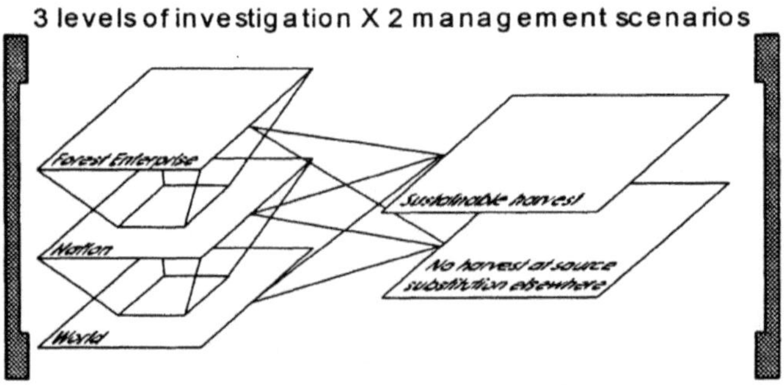

Figure 2. Layout for the investigation. The risks of two management scenarios (i.e. policy options) are analyzed for three geographical levels.

Risk analysis: a historical perspective and literature review

Risk analysis as a scientific discipline had its beginning at the turn of the 18th century, when catastrophic events were no longer considered simple acts of destiny, but as relations of causes and effects. Probably the first major event that was systematically analyzed on scientific grounds was the Lisbon earthquake of 1755 (Schweizer Rück, 1991). New technologies such as deep-mining or steam power opened the field for both retrospective and preventive analysis of risks in the 19th century, although the development of formal methodologies lagged behind considerably. The most advanced field of risk analysis was probably boiler technology, where frequent severe incidents soon drew the attention of both the regulatory and the technical community. Professional associations for boiler supervision stood at the origin of many standardization and regulatory bodies that were active in the later development of risk analysis methods (Minssen 1878). However, in these early years, the issue of safety and thus also the application of risk analysis was perceived to be primarily an engineering discipline.

The embedding of risk analysis in a broader concept of safety science is still relatively young. The introduction of new technologies with associated risks of unprecedented scale in the second half of the 20th century added the notions of risk evaluation and risk management to risk analysis. Nuclear weapons, nuclear power plants and large-scale non-nuclear industries awakened the concern of the public. Risk was no longer something to be handled solely by engineers, and the safety science as a whole moved from the technical into the socio-economic arena. This change is also reflected in the literature. The early risk-related literature focuses almost exclusively on risk analysis and technical means to improve the safety of individual components (see for example Minssen 1878, Meier 1992, SEV 1983). There is a limited number of early publications that discuss risks in a socioeconomic context (e.g. Culmann 1864), but the conclusions and recommendations still focus on technical details rather than on social or policy issues.

A milestone with regard to the development of the risk analysis methodology is the so-called Rasmussen Report compiled by the US Nuclear Regulatory Commission to assess the safety of commercial reactor types (NRC 1975). It is probably the first comprehensive application of systematic risk analysis for large technological systems. Similar studies have subsequently been undertaken in Western European countries. The methodology has also been transferred to other fields such as chemical processes and biohazards, and today risk analyses are requested by many legal procedures that apply to technological systems (such as the Swiss executive ordinance on major accidents StFV). A short summary of the most popular risk analysis methods used today as well as some software tools available is contained in Beroggi and Kröger (1993).

A relatively young application field for risk analysis are natural hazards. Although there is a long tradition of dealing with these hazards, their understanding in a safety science context has emerged only recently. Petak and Atkisson (1982) are among the first to introduce the notion of risk analysis by presenting case studies of historical hazards in a contemporary environment. Hollenstein (1997) evaluates the potential application of risk analysis methods to natural hazards. The author also participated in a multidisciplinary project aiming at the development of practical methodologies for analyzing natural risks (BUWAL 1998). An example for a modern application of risk analysis for gravitational hazards is presented in BUWAL (1999a, 1999b).

However, with regard to the risk analysis application that is suggested here, there is very little literature available. Using risk analysis as an independent tool for evaluating the effects of different policy options (as opposed to applying it for considering risk as one aspect in a policy setting process) is far from being a normal procedure. Hennen (1995) suggests that risk analysis should be part of the technology assessment, but without explicitly referring to it as an evaluation tool. Hessler (1995) mentions the importance of risk communication (and thus implicitly also risk analysis) within the policy finding process, but it is not obvious how such an integration should be performed. Briggs (1999) states that "...environmental health indicators [are] ... a means of ... helping to prioritise and rank ... issues as a basis for action"; risk analysis is a method of determining these indicators. Perhaps the strongest statement in favour of using risk analysis for policy evaluation is found in the conclusions and recommendations of a workshop for the National Environmental Health Action Plans (Briggs et al. 1999), where it is said that "risk assessment is a powerful tool ... for priority setting and for defining relevant actions ..." and that "risk assessment should be used as an integral part of the NEHAP process wherever practicable". Still, these applications deal with policy setting in the framework of man-made environmental risks. So far, risk analysis has obviously not been used on a large scale as a decision tool for planning resource management issues such as forest policy development.

Risk analysis is only one technique for evaluating and supporting decision-making for problems involving multiple criteria, i.e. an approach for multi-criteria decision analysis MCDA as defined e.g. by Lootsma (1999). Decision analysis as a field of methodological research addresses the whole spectrum of approaches that exists. Bunn (1984) characterizes decision analysis as a way to view "complex decision problem[s] ... not ... holistically, but as an assemblage of individual features", and including uncertainty, multiple objectives, multiple options and sequential stages as factors complicating the decision. Thus, decision analysis is a meta-methodology that needs to be implemented using an actual technique like risk analysis, cost-benefit analysis, utility theory, game theory (von Neumann and Morgenstern, 1944), analytic hierarchy process (Saaty 1980) or any other method that for making rational decisions (i.e. those based on objective procedures and presumably free of subjective bias, but not free of subjective preferences). Decision analysis is thus regarded as a key concept in the decision making process, although it has been criticited for having only technology perspective and lacking personal and organizational perspectives (Linstone, 1984). The rationale for using risk analysis instead of any other MCDA technique is its focus on avoiding negative outcomes rather than on maximizing utility. It is the author´s opinion that such a conservative approach is particularly well suited for decisions in forestry production that affect complex natural and socio-economic systems on a large scale and over a long period.

This view is also supported by the positioning of risk analysis in comparison to other decision support tools provided by Hofstetter (1998).

Methodology

In order to use risk analysis as a tool for policy evaluation, it is necessary to develop an analogy between these two applications such that they can be projected on each other, i.e. each module of one procedure must be contained in the other one as well. However, it is not required that they both have the same structure. If appropriate, it is possible to group or split individual modules.

Weimer and Vining (1992) suggest the following seven-step structurization for the policy analysis process:

1. Understanding the problem
2. Choosing and explaining relevant goals and constraints.
3. Choosing a solution method.
4. Choosing evaluation criteria.
5. Specifying policy alternatives.
6. Evaluating: predicting impacts of alternatives and valuing them in terms of criteria.
7. Recommending actions.

Risk analysis can be structured as follows (adapted from Hollenstein, 1997):

1. System definition and description.
2. Selection of risk variables.
3. Selection of an analysis technique.
4. Calculation of the risk for each combination of system and risk variable.
5. Aggregation of risks for each system.
6. Selection of optimal system.

Now the two procedures need to be mapped on each other (Fig. 3). Step 5 of the policy analysis is considered irrelevant in the present context, because the policy alternatives are the subject of the evaluation. The different available options are known beforehand and formulated as different systems in step 1 of the risk analysis structure. The relevant goals and the evaluation criteria of the policy analysis (steps 2 and 4) are combined in the risk variables (step 2). Step 6 of the policy analysis, the evaluation, is split up in two steps in the risk analysis: calculation and aggregation. Recommending actions (step 7 of the policy analysis) is comparable to the selection of the optimal system (i.e. in the present case the one with the lowest associated risk). The mapping is important to ensure that the substitution of the policy analysis methodology by the risk analysis does not result in an incomplete procedure. Now that this is shown, it is no longer necessary to formulate the task in terms of a policy analysis, but it can directly be defined as a risk analysis problem.

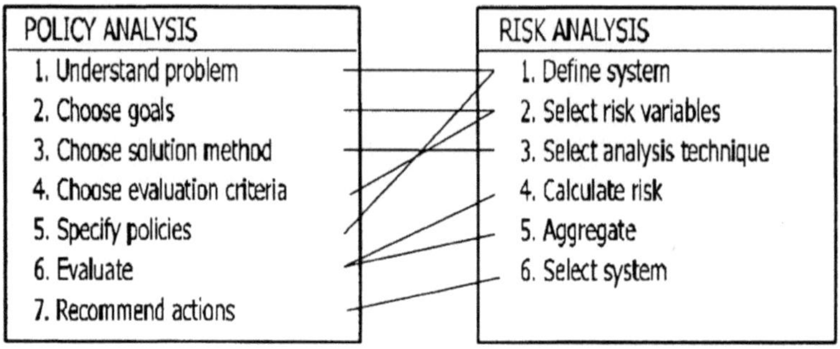

Figure 3. Matching activities between prescriptive policy analysis and risk analysis procedures. For every policy analysis step, an equivalent step within a risk analysis can be determined.

The system definition delimits the area under consideration as well as the boundaries and cross-boundary interactions that are risk-relevant. In the present example, the extension of the system depends on the level investigated. On the forest owner level it comprises only the forest and the inputs in the forest (in terms of production factors) and the output from the forest (in terms of products). On a national level, the forest area is a spatial subsystem of the whole system (i.e. the country), and the inputs into it and outputs from it are at least partly intrasystemic. On a global level, there are no more cross-boundary interactions, i.e. all the risk transfers happen within the system (Fig. 4).

The selection of risk variables is necessary for the calculation and aggregation steps. Risk variables may be defined as ways to measure the negative effects that can result from a process. To make sure that the analysis results for the different levels are comparable, the same set of risk variables must be applied to all the systems under consideration. There is no "scientific" way to determine what relevant variables are, since they depend on individual preferences and values. For the purposes of this study, the following risk variables are selected:

1. Loss of human life (number of fatalities)
2. Disturbances of human health (number of persons*time)
3. Loss of ecological functionality (area)
4. Loss of soil fertility (% of net primary production)
5. Net emission of greenhouse gases (mass)
6. Financial losses

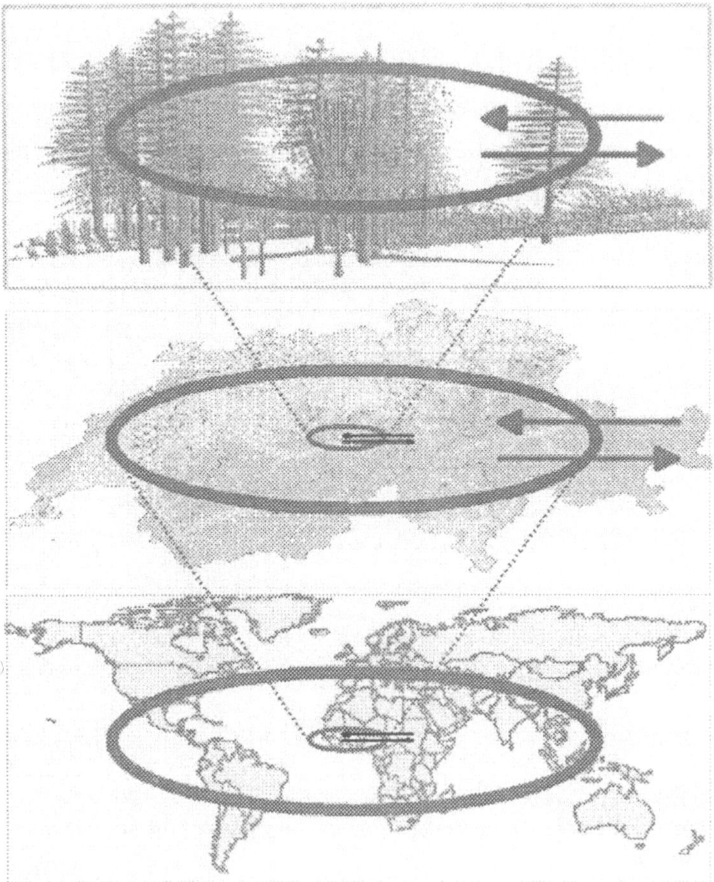

Figure 4. System boundaries and transboundary input / output relations. Effects that are external on the ownership level may be internal already on a national or at least on a global level.

Some of the variables are rather vague and fuzzy, e.g. the one for loss of ecological functionality. This is not too critical for this special application of risk analysis, since the relative ranking of options is of interest rather than an absolute risk number, and in most cases it will be possible to determine which options have a higher risk for a certain variable. However, the fuzziness complicates the calculation considerably as soon as a trade-off between different variables is introduced (i.e. in the aggregation step).

The selection of a risk analysis technique usually depends on the system under consideration, and it not possible to indicate a single, most suitable method. It is important that the method selected reflects the accuracy of the available data, i.e. highly detailed methods should not be used in a case like this where the knowledge about the system is at best limited. Furthermore, the study focuses on the use of the concept of risk analysis for policy evaluation rather than on specific analysis methods. Thus, the risk associated with a certain system is represented using a simple product of the frequency of occurrence and the intensity of the consequences of a negative event for each risk variable. Such a relation can be illustrated in a so-called frequency-number diagram (Fig. 5).

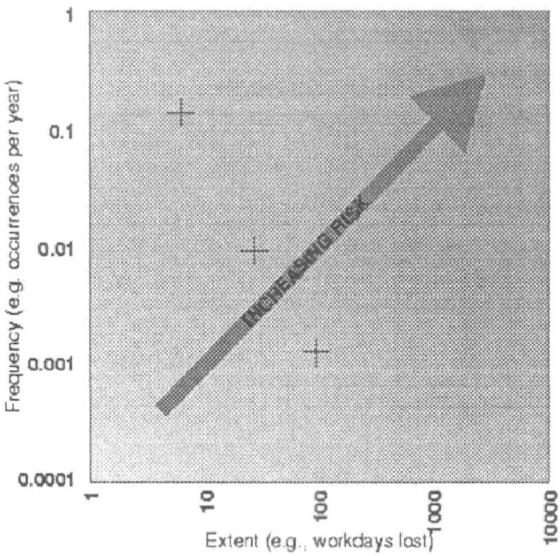

Figure 5. Frequency-Number diagram (F/N diagram). Based on the definition of risk, a negative event can be characterized using its consequences (number) and its probability of occurrence.

The calculation of the risk for each risk variable itself is straightforward. Whenever a risk variable is relevant for more than one possible event within the same system (e.g. fatalities due to logging and due to transportation), the risks for the events are added up, provided the events are statistically independent. This calculation scheme produces linear risk estimates, i.e. a calculatory risk aversion is not introduced. Risk aversion factors are sometimes used to account for the nonlinear perception that catastrophic accidents have. However, it is assumed that none of the events investigated herein reach catastrophic extents. Furthermore, it is the author's opinion that risk aversion has to be accounted for in the risk evaluation process (see Hollenstein 1997) rather than in the risk analysis.

The aggregation of the total risk for a system over all risk variables requires trade-off relations between the different scales. These relations are obviously subjective, and they depend on individual and social preferences, i.e. on the cultural environment in which the analysis is performed. For the purposes of the study, the trade-offs are selected arbitrarily; where possible, the relations used in the manual for the Swiss executive ordinance on major accidents (Handbuch zur Störfallverordnung, BUWAL 1991, 1992) are applied after necessary adaptations.

The trade-off relations are crucial for the result of the risk analysis, and therefore also for the policy making. It is the responsibility of the respective decision-making entity (e.g. forest owner, lawmakers) to determine the trade-offs if the concept outlined in this article is used in real-world applications, and it is more than likely that this is only possible after intense discussions.

In principle, it is possible to stop the risk analysis short of the aggregation step, and the different systems considered (i.e. policy options available) can be evaluated using a pairwise comparison for each risk variable. However, a system selected based on this approach can only be proven to be optimal if it is superior to all the others for all the variables investigated. Such a case is considered trivial (elimination of dominated solutions) and therefore not discussed any further. As soon as every system is inferior or superior to another one with regard to different variables, optimization requires aggregation.

The procedure for selecting the optimal system with the calculation method outlined above is again straightforward. Since there are no nonlinearities and since the (intuitive and often hardly reproducible) step of comparing totally different effects is made transparent, the optimal choice is the system with the lowest aggregated risk. Any other selection is not justifiable unless additional aspects are included that extend beyond the "engineering" approach of using risk analysis for policy evaluation.

As mentioned, a modern safety science approach includes risk evaluation and risk management aspects as well as risk analyses. Neither risk evaluation nor risk management is explicitly addressed within the present study, but they are implicitly part of the process. It is assumed that the risk analysis is performed by an entity with decision-making competence (e.g. the forest owner), and the decision-maker´s risk evaluation is expressed in the selection of trade-off relations, where different values are related to each other. Risk management is dealt with in a similar way, by defining the options under consideration and the restrictions that apply to them. This allows for example to define options based on new technologies that have different risk profiles and thus score differently in the analysis.

The issues of risk evaluation and management are considerably more difficult to address if the risk analysis is not performed by, but externally provided for a decision-making entity. In such a case, the value systems of the analyzing and the decision-making entity may differ, and a mechanism for dealing with these differences must be established. However, this is beyond the scope of the present study.

Data Requirements, Data and Assumptions

First the data requirements for the risk analysis must be determined. This is done in three steps (Fig. 6):

1. Definition of subsystems, operations and system states of interest,
2. definition of activities or processes associated with the operations or states,
3. definition of parameters that are required to assess the effects of the process on the risk variables.

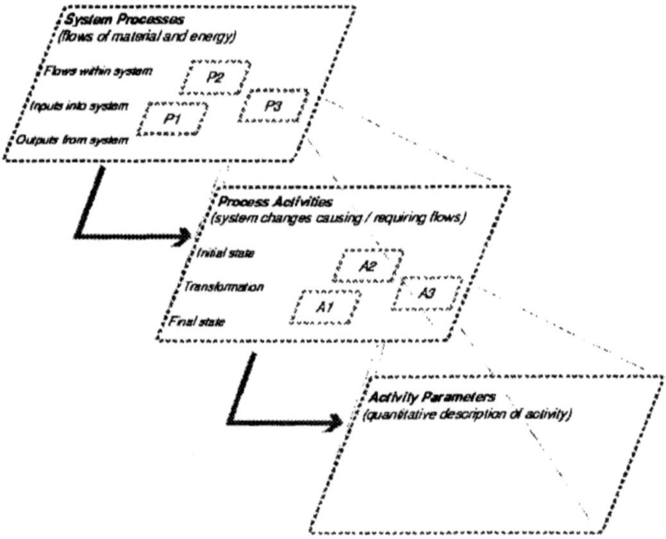

Figure 6. Determination of data requirements. Starting with the operational states of the system, a set of potentially risk-generating processes can be defined. Subsequently, the parameters required for assessing the risk can be determined.

According to the policy scenarios mentioned in the introduction, two system states can be defined for each system considered:

1. Normal operation, i.e. a more or less steady state with temporally and spatially stable and continuous pattern,
2. operation after a catastrophic event, i.e. a temporally and perhaps also spatially unstable transient pattern.

For the management scenario with sustainable wood harvest, the normal operation in the forest consists of the biological production subsystem [P_{Bio}] that increases the biomass stock at a rate P^*_{Bio} (as expressed for example by the Net Primary Production NPP), and the human consumption subsystem [C_{Bio}] that decreases it at rate C^*_{Bio} (i.e. the amount of biomass harvested within a certain period). Sustainability requires that

$$P^*_{Bio} - C^*_{Bio} \geq 0 \qquad [1]$$

where P^*_{Bio} and C^*_{Bio} are averages of the biomass flow over time and space. Thus, the growth is the upper limit for the consumption, and this is also the most active normal state of the system.

Assuming that for mid-European forests, wood is the only type of biomass that is extracted in significant amounts, this equation can be rewritten using only the woody parts P^*_{Wood} and C^*_{Wood} such that:

$$P^*_{Bio} - P^*_{Wood} \geq 0 \qquad [2]$$

and

$$C^*_{Bio} - C^*_{Wood} \geq 0 \qquad [3]$$

and

$$P^*_{Wood} - C^*_{Wood} \geq 0 \qquad [4]$$

It can be assumed that a specific positive risk is associated with the production and consumption of a unit of biomass or wood under certain conditions. The amount of this risk depends on various factors (e.g. technology, environment). Based on this assumption, the following case is of particular interest:

$$P^*_{Wood} - C^*_{Wood} = 0 \qquad [5]$$

Equation [5] describes the condition under which the risks associated with the wood harvest are at maximum.

For the scenario without any biomass harvest, the decrease of the biomass stock in a steady state is due to natural processes, i.e.

$$C^*_{Bio} = 0 \qquad [6]$$

and therefore also

$$C^*_{Wood} = 0 \qquad [7]$$

Extending the analysis beyond the borders of the forest also extends the spectrum of processes that require consideration within the operation. Apart from the biological production subsystem, other subsystems that are related to the subsequent handling of the (woody) biomass need to be included into the analysis. Among them, the ones covering transportation, $[Tr_{Bio}]$ and $[Tr_{Wood}]$, and processing or utilization activities, $[Ut_{Bio}]$ and $[Ut_{Wood}]$, are of primary interest. Associated with these subsystems are flows Tr^*_{Bio}, Tr^*_{Wood}, Ut^*_{Bio} and Ut^*_{Wood}. Moving the analysis from the national to the global level does physically extend the transportation and utilization subsystems, but it does not add any new subsystems or processes within the analysis framework (Fig. 7). As before, the extremes of the flows are of interest for the analysis:

$$Tr^*_{Bio} = C^*_{Bio} \qquad [8]$$

$$Tr^*_{Wood} = C^*_{Wood} \qquad [9]$$

$$Ut^*_{Bio} = C^*_{Bio} \qquad [10]$$

$$Ut^*_{Wood} = C^*_{Wood} \qquad [11]$$

for the scenario with sustainable harvest and

$$Tr^*_{Bio} = 0 \qquad [12]$$

$$Tr^*_{Wood} = 0 \qquad [13]$$

$$Ut^*_{Bio} = 0 \qquad [14]$$

$$Ut^*_{Wood} = 0 \qquad [15]$$

for the scenario without biomass removal.

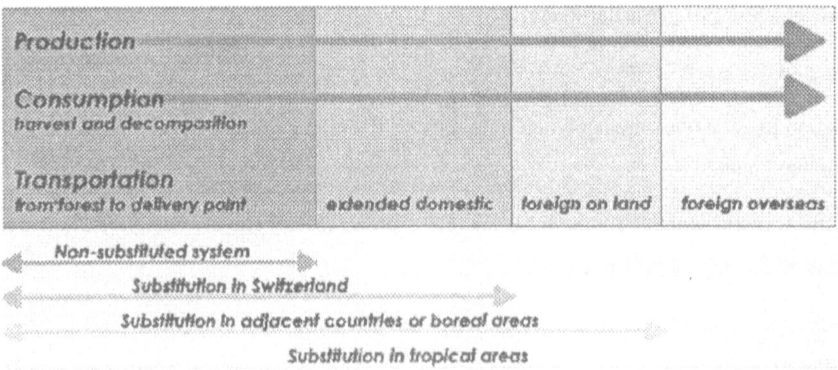

Figure 7. Transferring the analysis from the national to the global level: Extension of the transportation subsystem.

Now it is possible to perform the third step in determining the data requirements, that is, to define parameters that can be used to assess the effects of the processes on the different risk variables. Table 1 shows a list of the parameters required for the analysis together with the risk variables they affect.

	Operational state			
Process (Flows)	***OI (normal operation)***		***OII (operation after catastrophe)***	
Biological production (P^*_{Wood})	Growth	Changes in fertility Greenhouse gas fixation Financial losses (gains)	Growth	Changes in fertility Greenhouse gas fixation Financial losses (gains)
	Composition	Loss of ecofunctionality[I] Changes in fertility[I] Financial losses (gains)	Composition	Loss of ecofunctionality[I] Changes in fertility[I] Financial losses (gains)
			Resilience	Loss of ecofunctionality Changes in fertility Greenhouse gas emission
Human consumption (C^*_{Wood})	Wood quantity	Loss of life Health impacts Loss of ecofunctionality Changes in fertility Greenhouse gas emission Financial losses (gains)	Wood quantity	Loss of life Health impacts Loss of ecofunctionality Changes in fertility Greenhouse gas emission Financial losses (gains)
	Accident rate	Loss of life Health impacts Loss of ecofunctionality[II] Changes in fertility[II] Financial losses	Accident rate	Loss of life Health impacts Loss of ecofunctionality[II] Changes in fertility[II] Financial losses
	Fatality rate	Loss of life	Fatality rate	Loss of life
	Area affected	Loss of ecofunctionality Changes in fertility	Area affected	Loss of ecofunctionality Changes in fertility
	Impact intensity	Loss of ecofunctionality Changes in fertility	Impact intensity	Loss of ecofunctionality Changes in fertility
	Impact duration	Loss of ecofunctionality Changes in fertility	Impact duration	Loss of ecofunctionality Changes in fertility
	Productivity	Financial losses (gains)	Productivity	Financial losses (gains)
	System cost	Financial losses	System cost	Financial losses
	Product prices	Financial gains	Product prices	Financial gains
	Fuel use	Greenhouse gas emission	Fuel use	Greenhouse gas emission
			Wood usability	Financial losses (gains)
			Type of storage	Loss of ecofunctionality Changes in fertility Financial losses
			Cost of storage	Financial losses
Transport (Tr^*_{Wood})	Distance	Loss of life Health impacts Financial losses (gains)	Distance	Loss of life Health impacts Financial losses (gains)
	Accident rate	Loss of life Health impacts Financial losses	Accident rate	Loss of life Health impacts Financial losses
	Fatality rate	Loss of life	Fatality rate	Loss of life
	Payload	Loss of life Health impacts Financial losses	Payload	Loss of life Health impacts Financial losses
	System cost	Financial losses	System cost	Financial losses
	Fuel use	Greenhouse gas emission	Fuel use	Greenhouse gas emission
Utilization (Ut^*_{Wood})	Purpose	Loss of life Health impacts Greenhouse gas emission Financial losses (gains)	Purpose	Loss of life Health impacts Greenhouse gas emission Financial losses (gains)
	Technology	Loss of life Health impacts Loss of ecofunctionality Changes in fertility Greenhouse gas emission Financial losses (gains)	Technology	Loss of life Health impacts Loss of ecofunctionality Changes in fertility Greenhouse gas emission Financial losses (gains)

Table 1. Parameters required for the analysis and risk variables affected.

[I] Relevant only if composition is altered by the effect of subsystem C^*_{Wood}. [II] Relevant only if environmental effects of accident are considered.

The relations shown in Table 1 cover only the direct effects of the system processes, i.e. no proportional share in the effect of commonalties is taken into account. It is for example assumed that the transportation subsystem is completely based on existing roads and that an increase in the amount of wood transported does not require additional roads and thus not result in loss of ecofunctionality or fertility (associated marginal risk equals zero). This is not totally correct, but with the available information, it is not possible to assign a share in the effects of commonalties.

Table 1 could be refined considerably by splitting up the four subsystems into a much more detailed structure for the calculation. However, since the goal of the study is a comparison of different policies rather than deriving absolute risk numbers for individual solutions, it is important that the same depth of analysis is applied to all the options. This can only be achieved with a high-level analysis due to the different nature and extent of the systems that are compared to each other in the subsequent investigation.

Table 2 contains the data that are used for the actual calculation of the risk associated with the two policies (harvest and no harvest) and the two scenarios (normal operation and operation after catastrophic event). It is important to remember that these data cover only the risks related to the policy options of the forest owner, i.e. only the consequences that follow immediately from harvesting or not harvesting wood in the forest considered are taken into account, but not the risks caused by substituting the production elsewhere.

To properly analyze the risks associated with the utilization subsystem, it is necessary to make an assumption as to where and how the wood is processed. While data about the distribution of the wood consumption for different end uses (referred to as purpose in Table 1) is available, it was not possible to collect the respective risk relevant data (technology) within the framework of this study. Therefore, the investigation is limited to raw logs delivered to the point of further processing in Switzerland (e.g. a sawmill, a pulp or paper factory).

	Parameter	*Source*	*SI / OI*	*SI / OII*	*SII / OI*	*SII / OII*
	Area	assumption	100 ha	100 ha	100 ha	100 ha
[P_{wood}]	Growth	assumption	10 $m^3ha^{-1}yr^{-1}$	10 $m^3ha^{-1}yr^{-1}$	10 $m^3ha^{-1}yr^{-1}$	10 $m^3ha^{-1}yr^{-1}$
[P_{wood}]	Composition	assumption	avg. value[I]	avg. value[I]	avg. value[I]	avg. value[I]
	Wood quantity	[P_{Wood}][II] assumption	1000 m^3	5000 m^3	10 m^3	100 m^3
	Accident rate (logging)	Rehschuh et al. (1984), Löffler et al (1994).	$2*10^{-4}h^{-1}$	$3*10^{-4}h^{-1}$	$2*10^{-4}h^{-1}$	$3*10^{-4}h^{-1}$
	Expected duration of disablement	assumption[III]	5 d	5 d	5 d	5 d
	Fatality rate	Butora (1978), assumption[IV]	1 %	2 %	1 %	2 %
	Healthcare cost	assumption	Fr. 5000.--	Fr. 5000.--	Fr. 5000.--	Fr. 5000.--
	Area cleared by final harvest[V]	assumption	1 ha	10 ha	0 ha	0 ha
	Area thinned		10 ha	0 ha[VI]	0 ha	0 ha
	Area passed by vehicles[VII]	Williamson 1990 Senyk et al. 1997	20 %	20 %	0 %	0 %
	Area with soil compaction[VII]	Senyk et al. 1997	5 %	5 %	0 %	0 %
	Impact intensity final harvest[VIII]	assumption	10 %	20 %	0 %	0 %
	Impact intensity thinning	assumption	5 %	n/a	0 %	0 %
	Impact intensity vehicle pass[IX]	Froehlich et al. 1986	20 %	20 %	0 %	0 %
	Impact duration harvest / thinning	assumption	1 yr	1 yr	0 yr	0 yr
	Impact duration vehicle pass	Williamson 1990, Froehlich et al. 1986	3 yr	5 yr	0 yr	0 yr
	Productivity of logging	assumption[X]	1 m^3h^{-1}	1 m^3h^{-1}	1 m^3h^{-1}	1 m^3h^{-1}
	Fuel use	Knechtle 1997, Winkler 1997	2.5 kgm^{-3}	2.5 kgm^{-3}	0.7 kgm^{-3}	0.7 kgm^{-3}
	System cost (logging)	assumption	60 $sFrh^{-1}$	60 $sFrh^{-1}$	60 $sFrh^{-1}$	60 $sFrh^{-1}$
	Wood usability	assumption	n/a	60 %	n/a	0 %
[C_{wood}]	Type of storage	assumption	n/a	irrigated[XI]	n/a	n/a
[C_{wood}]	Storage duration	assumption	n/a	3 yr	n/a	n/a
[C_{wood}]	Cost of storage	FIO 1993	n/a	6.5 $sFrm^{-3}yr^{-1}$	n/a	n/a
	Domestic transport[XII]	assumption	100 km (one way)	100 km (one way)	n/a	n/a
	Accident rate (transport)	BFS 1998	10^{-6} km^{-1}	10^{-6} km^{-1}	n/a	n/a
	Serious injury rate	BFS 1998	$7.5*10^{-2}$	$7.5*10^{-2}$	n/a	n/a
	Expected duration of disablement	assumption[III]	5 d	5 d	5 d	5 d
	Fatality rate	BFS 1998	$7.5*10^{-3}$	$7.5*10^{-3}$	n/a	n/a
	Healthcare cost	assumption	5000 sFr	5000 sFr	n/a	n/a
	Repair cost	BFS 1998	6500 sFr	6500 sFr	n/a	n/a
	Payload	assumption	10 t	10 t	n/a	n/a
[Tr_{wood}]	Fuel use	Winkler 1997, assumption	0.05 $kgt^{-1}km^{-1}$	0.05 $kgt^{-1}km^{-1}$	n/a	n/a
[Tr_{wood}]	System cost (transport)	assumption	2 $sFrkm^{-1}$	2 $sFrkm^{-1}$	n/a	n/a

Table 2. Data used for the risk analysis calculation for the two policy options and the two scenarios for the forest in Switzerland. No risks from substituting the production are taken into account.

[I] Only the commercial value of the composition is relevant (for assessing the financial losses and gains), since no alteration by the subsystem is assumed. [II] The volume for normal operation is derived from the growth figures using the scenario harvest and no harvest, respectively; for the operation after catastrophe, a fivefold amount of the annual growth is assumed to be affected. A limited amount of wood needs to be worked on for safety reasons (opening trails etc.) even in the no harvest scenario, but this wood is not extracted from the forest. [III] Löffler et al. (1994) and Rehschuh et al. (1984) indicate average durations of disablement that are higher, but their studies include a high percentage of part-time loggers with generally more severe accidents. [IV] Based on experience by the Swiss Accident Insurance Agency SUVA that on average, one fatality occurs for every 500'000 m^3 logged under normal operation and for every 150'000 m^3 after storm events (L. Rigling 1990, personal communication). [V] Cleared refers to a complete removal of the trees and subsequent reforestation. [VI] Assuming that no thinning is performed in the immediate aftermath of a catastrophe due to a lack of resources. [VII] Reference: area logged or thinned. [VIII] Expressed as a reduction of the ecosystem quality. [IX] Expressed as a reduction of the soil fertility. [X] The productivity is relevant mainly for the estimation of accidents; therefore the felling process as the most accident-prone is of particular interest. The figure used here corresponds to empiric values measured in Switzerland. [XI] Irrigated storage is assumed to affect neither the ecological functionality nor the fertility of the site; i.e. the storage type is only relevant with regard to its cost. [XII] For the scenario harvest, the transport refers to the harvested wood; for the no harvest scenario, it refers to the substitute wood.

As can be seen from Table 2, only point estimates are available for most of the variables needed for the calculation, i.e. only one point in a F/N diagram (cf. Fig. 5). These point estimates represent mean, maximum likelihood or median values. It is obvious that for a thorough analysis, much more information about the distribution of the variables, especially their variability, is desirable. However, additional statistical information is only available in very few cases (e.g. for the soil impact data), and these rare occurrences do not legitimate an overall error calculation.

Realistically, one must assume that a reduction in the net amount of wood produced does not significantly affect the consumption of wood products in a highly developed country such as Switzerland. In spite of its bulky nature, wood is easily transported over very long distances, and the domestic demand is therefore potentially covered by supplies from all over the world. If the policy options are analyzed on a national or global level, the risks of substitute production systems can no longer be neglected to avoid an unbiased picture. For an industrial mass product like wood, one of the main factors for selecting a supply source is the price at which the goods can be delivered. For the present study, it is simply assumed that for every unit wood not harvested in the forest considered in Switzerland, the same amount needs to be harvested elsewhere, most likely in a place with a low direct product cost. The risks of this production, together with the increased transport risk (longer distance) must then be added to the no harvest scenario to provide a useful comparison.

The risk associated with a substitute production system presumably depends on its socio-economic and technological environment. It is therefore necessary to make some assumptions as to where the replacement wood will come from, and five possibilities are of particular interest:

1. Substitution by other forests in Switzerland, i.e. from within the same socio-economic and technological system;

2. substitution by forests in neighbouring countries (Germany, France, Austria) with a similar socio-economic and technological setting;
3. substitution by boreal forests managed under a highly mechanized harvesting regime, such as found in Sweden, Finland or Canada, with high capital and low manpower input;
4. substitution by boreal forests managed under a harvesting regime with limited mechanization, such as found in countries of the former Soviet Union, with low capital and high manpower input;
5. substitution by tropical forests managed under a regime with very limited mechanization, with very low capital and very high manpower input.

For the first and second of these five possibilities, the same raw data apply as used in Table 2 (with the exception of the transport distance and, for neighbouring countries, also the transport accident rate) because of their similarities with the non-substituted production system.

For the remaining three possibilities, all the raw data are subject to variations because of differences both in the growth conditions and the management concepts. Table 3 contains the data that are relevant for the two boreal and the tropical substitute systems that were investigated on the national and global risk analysis level. Only the data for the sustainable harvest policy option are provided since the substitutes are only considered under a production point of view. Also, only normal operations are considered for the substitute systems, since it is not likely that they are subject to catastrophic impacts exactly at the same time as the original production system. Therefore, the substitution needs only to account for the amount of wood that would be usable in the non-substituted production.

	Parameter	Source	Boreal, highly mech.	Boreal, low mech.	Tropical
	Land base	assumption	200 ha	200 ha	1000 ha
	Growth	assumption Brouwer 1996[I]	5 $m^3ha^{-1}yr^{-1}$	5 $m^3ha^{-1}yr^{-1}$	1 $m^3ha^{-1}yr^{-1}$
[P_{wood}]	Composition	assumption	avg. value[I]	avg. value[I]	avg. value[I]
	Wood quantity	substituted system (usable wood only)[II]	1000 m^3 / 3000 m^3	1000 m^3 / 3000 m^3	1000 m^3 / 3000 m^3
	Accident rate (logging)	Rehschuh et al. 1984, Löffler et al 1994., assumption[III]	$1.2*10^{-4}h^{-1}$	$2*10^{-4}h^{-1}$	$3*10^{-4}h^{-1}$
	Expected duration of disablement	assumption	5 d	5 d	5 d
	Fatality rate	assumption[IV]	0.5 %	1 %	2 %
	Healthcare cost	assumption	5000 sFr	5000 sFr	5000 sFr
	Area cleared by final harvest	assumption	2 ha	n/a[V]	n/a[V]
	Area thinned	assumption	20 ha	n/a[V]	n/a[V]
	Area affected by logging	assumption	n/a	10 ha	10 ha
	Area passed by vehicles[VI]	Williamson 1990 Senyk et al. 1997	20 %	40 %	40 %
	Area with soil damage[VII]	Senyk et al. 1997	5 %	10 %	10 %
	Impact intensity final harvest	assumption Jonkers 1987[VIII]	10 %	20 %	30 %
	Impact intensity thinning	assumption	5 %	n/a	n/a
	Impact intensity vehicle pass	Froehlich et al. 1986, assumption	20 %	30 %	30 %
	Impact duration logging	assumption[IX]	2 yr	2 yr	1 yr
	Impact duration vehicle pass	assumption[X], Williamson 1990	5 yr	5 yr	3 yr
	Productivity of logging	assumption[XI]	2.5 m^3h^{-1}	1 m^3h^{-1}	0.5 m^3h^{-1}
	Fuel use	Knechtle 1997, Winkler 1997[XII]	2.5 kgm^{-3}	2.5 kgm^{-3}	2.5 kgm^{-3}
[C_{wood}]	System cost (logging)	assumption	100 $sFrh^{-1}$	30 $sFrh^{-1}$	30 $sFrh^{-1}$
	Transport distance	assumption	2000 km (one way)	3000 km (one way)	10000 km (one way)[XIII]
	Accident rate (transport)	BFS 1998, assumption	$1.5*10^{-6}$ km^{-1}	$2*10^{-6}$ km^{-1}	10^{-6} km^{-1}[XIV]
	Serious injury rate	BFS 1998	$7.5*10^{-2}$	$7.5*10^{-2}$	$7.5*10^{-2}$
	Expected duration of disablement	assumption	5 d	5 d	5 d
	Fatality rate	BFS 1998	$7.5*10^{-3}$	$7.5*10^{-3}$	$7.5*10^{-3}$
	Healthcare cost	assumption	5000 sFr	5000 sFr	5000 sFr
	Repair cost	BFS 1998	6500 sFr	6500 sFr	6500 sFr
	Payload	assumption	10 t	10 t	10 t
[Tr_{wood}]	Fuel use	Winkler 1997, assumption	0.05 $kgt^{-1}km^{-1}$	0.05 $kgt^{-1}km^{-1}$	0.025 $kgt^{-1}km^{-1}$
	System cost	assumption	0.5 $sFrkm^{-1}$	0.3 $sFrkm^{-1}$	0.1 $sFrkm^{-1}$

Table 3. Raw data for the substitute production systems.

[I] Assuming that a sustainable, i.e. non-degrading management system is applied, nutrient cycling limits the yield in the tropics (low net primary production NPP) [II] The wood quantity to substitute is derived for both normal and catastrophic operation of the original production system. [III] Mechanization has led to a decline of the accident frequency of about 30% in Sweden. On the other hand, due to increased exposure time, an increase in accidents can be expected when the percentage of non-mechanized work increases. [IV] The frequency of fatal accidents is assumed to increase with increasing exposure of people. [V] It is assumed that both the low mechanized systems do not apply thinnings, but only harvesting operations. [VI] The area passed by vehicles increases with decreasing yield per unit surface, decreasing degree of procedural sophistication and increasing precipitation (tropics). [VII] Besides compaction, soil erosion is also relevant both for boreal and tropical sites. [VIII] Because of the much higher number of tree species per ha, the redundancy of biotopes is generally lower; i.e., disturbances are more disruptive especially for species with limited mobility. [IX] Lower biological activity in boreal forests presumably delays recovery of harvested site. [X] Again, a delayed recovery is assumed. [XI] The productivity in the tropics is assumed lower due to the lower intensity per unit surface (low NPP). [XII] Based on the energy input, the effect of the mechanization on fuel use is not significant. [XIII] It is assumed that a significant part of the transport from the tropics to Switzerland is done by vessel, and that only 1000 km distance are covered by road transport with the truck returning empty. [XIV] The overall accident rate is assumed lower due to the transport by vessel.

Using the data in Table 2 and Table 3, it is possible to generate point estimates of the risk associated with the processes described in Table 1 and to evaluate the effect of substituting domestic by foreign wood production for each risk variable. In some cases, this may already be sufficient for a comparison, but usually an aggregation is still required such that different options can be tested against each other using a single risk value. The trade-off relations used for the aggregation process are shown in Table 4. The basis for determining these relations are the scales contained in the Appendix G of the Manual to the Swiss executive ordinance on major accidents (Handbuch zur Störfallverordnung StFV). There, risk indicators for the determination of an accident magnitude are given. With the dimensionless extent parameter used there, the following equations are derived:

$$e = 0.124 * \ln(n_{fat}) + 0.02773 \qquad [16]$$

$$e = 0.134 * \ln(n_{inj}) - 0.29255 \qquad [17]$$

$$e = 0.087 * \ln(n_{eco}) + 0.2 \qquad [18]$$

$$e = 0.091 * \ln(n_{sol}) + 0.2637 \qquad [19]$$

$$e = 0.134 * \ln(n_{fin}) - 0.2 \qquad [20]$$

with e = dimensionless accident extent, n_{fat} = number of fatalities, n_{inj} = number of injured people, n_{eco} = area of destroyed ecosystem in km^2, n_{sol} = area of contaminated soil in km^2, n_{fin} = discounted financial losses in million sFr.

For the greenhouse gas emissions, no such data are available, and the following approach is selected. It is assumed that the price of fossil fuel does only cover one third of its total cost (including all externalities). Assuming that the price of one litre diesel is sFr. 1.22 and accounts for 3 kg CO2, the greenhouse gas emissions relate to the financial losses like 1.22 $sFr*kg^{-1}$.

	Human life Fatalities/yr	Human health Persondays/yr[1]	Ecofunctionality ha	Soil fertility %	Gr. gas emissions kg/yr	Financial losses sFr/yr
Human life Fatalities/yr	$1.00*10^0$	$5.50*10^1$	$1.40*10^1$	$7.50*10^0$	$4.50*10^6$	$5.50*10^6$
Human health Persondays/yr	$1.82*10^1$	$1.00*10^0$	$2.54*10^1$	$1.36*10^1$	$8.18*10^4$	1.0010^5
Ecofunctionality ha	$7.14*10^2$	$3.93*10^0$	$1.00*10^0$	$5.36*10^1$	$3.21*10^5$	$3.93*10^5$
Soil fertility %	$1.33*10^1$	$7.33*10^0$	$1.87*10^0$	$1.00*10^0$	$6.00*10^5$	$7.33*10^5$
Gr. gas emissions kg/yr	$2.22*10^7$	$1.22*10^5$	$3.11*10^6$	$1.67*10^6$	$1.00*10^0$	$1.22*10^0$
Financial losses sFr/yr	$1.82*10^7$	$1.00*10^5$	$2.54*10^6$	$1.36*10^6$	$8.18*10^1$	$1.00*10^0$

Table 4. Trade-off relations between different risk variables used for the risk aggregation procedure.

[1] Assuming that the average duration of an injury is five days.

The values in Table 4 can now be used to compute an overall risk measure, called hereafter risk equivalent value REV, according to the formula

$$REV = n'_{fat} + 1.82*10^{-2}*n'_{inj} + 7.14*10^{-2}*n'_{eco} + 1.33*10^{-1}*n'_{sol} + 2.22*10^{-7}*n'_{gge} + 1.82*10^{-7}*n'_{fin} \quad [21]$$

where n'_{fat} = number of fatalities per year, n'_{inj} = number of persondays of injury per year, n'_{eco} = area affected by loss of ecological functionality (ha), n'_{sol} = loss of soil fertility (%), n'_{gge} = greenhouse gas emission ($kg*yr^{-1}$), n'_{fin} = financial loss ($sFr*yr^{-1}$).

Calculations

This section shows the calculations used to derive the point estimates for the individual risk variables and system activities. Most of these calculations are straightforward multiplications, additional explanations are given wherever necessary.

Biological Production

Loss of soil fertility

$$lsf(G) = (\Delta T * \Delta V_T * \Delta n_V) \quad [22]$$

$$lsf(C) = (\Delta T * ndp(C)) \quad [23]$$

$$lsf(R) = (nlr * \Delta T_R) \quad [24]$$

where lsf(G) = growth induced loss in soil fertility (%), ΔT = remaining growth cycle time or observation period (yr), ΔV_T = change in volume per unit of time ($m^{3*}yr^{-1}$), Δn_V = soil nutrient availability change per unit volume wood produced ($\%*m^{-3}$), $lsf_T(C)$ = composition-induced fertility loss (%), ndp(C) = nutrient depletion potential of a species composition ($\%*yr^{-1}$), lsf(R) = resilience-induced loss in fertility, nlr = nutrient leaching rate without vegetation cover ($\%yr^{-1}$), ΔT_R = lag time until recovery of vegetation stops leaching (yr). Sustainable production requires that the stock of available soil nutrients is not depleted below a critical value where production is affected; i.e. $\Delta n_V = 0$ for practical purposes in this study. Whereas positive influences of certain compositions (nitrogen fixating plants) on the soil fertility are documented, similar proof for negative effects remains scarce (Morris et al. 1994). Therefore, ndp(C) = 0 is assumed. Leaching after loss of vegetation cover is frequently observed on sites with soils having a limited cation exchange capacity (Bouwer 1996). However, this is generally not the case for the soils in Switzerland; and since a significant reduction of the fertility is not to be expected, nlr = 0 is assumed.

Greenhouse gas emission (CO_2 only)

The greenhouse gas emission is a negative value, i.e. the growth leads to fixation:

$$gge_T(G) = -(C_V * \Delta V_T) \qquad [25]$$

with $gge_T(G)$ = greenhouse gas (CO_2) emission per time (kg), C_V = specific content of greenhouse gas per unit of volume (kgm^{-3}). For reasons of simplicity, the specific CO_2 content for wood is assumed as being a constant with the value 925 kgm^{-3} (Winkler et al. 1997), even though, to be correct, the biomass would be a better reference.

Financial losses

The biological production results in the buildup of merchantable goods, and this can be considered as a financial gain or a negative financial loss. However, this gain can only be realized if the products are removed from the forest, and it is therefore considered within the consumption subsystem.

Human Consumption

The risk associated with the human consumption subsystem as defined here are all related to the activity of harvesting biomass.

Loss of life

$$lhl_T(H) = (\Delta V_T * prod_{HS}^{-1} * f_{AccHS} * fr_{HS}) \qquad [26]$$

with $lhl_T(H)$ = harvest-induced loss of human life per unit time (lifes*yr^{-1}), $prod_{HS}$ = productivity of the harvesting system considered (m^3h^{-1}), f_{AccHS} = accident frequency of the harvesting system used (h^{-1}), fr_{HS} = fatality rate of the harvesting system (%).

Disturbance of human health

$$dhh_T(H) = (\Delta V_T * prod_{HS}^{-1} * f_{AccHS} * d_{AccHS}) \quad [27]$$

where $dhh_T(H)$ = harvest-induced disturbance of human health per unit of time (person*days*yr^{-1}), d_{AccHS} = average duration of disablement due to harvesting accidents.

Loss of ecological functionality

$$lef_T(H) = (I_{impFH} * \Delta T_{impFH} * A_{impFH}) + (I_{impTh} * \Delta T_{impTh} * A_{impTh}) \quad [28]$$

with $lef_T(H)$ = harvest-induced loss of ecological functionality per unit of time (ha), I_{impFH} = intensity of final harvesting impact (%), ΔT_{impFH} = effective duration of final harvesting impact (yr), A_{impFH} = area affected by final harvesting impact (ha*yr^{-1}), I_{impTh} = intensity of thinning impact (%), ΔT_{impTh} = effective duration of thinning impact (yr), A_{impTh} = area affected by thinning impact (ha*yr^{-1}).

Loss of soil fertility

$$lsf(H) = (A_{SC} * I_{impSC} * \Delta T_{impSC}) \quad [29]$$

where $lsf(H)$ = harvest-induced loss of soil fertility (%), A_{SC} = relative area affected by soil compaction after vehicle passes (%*yr^{-1}), I_{impSC} = intensity of soil compaction impact (%), ΔT_{impSC} = effective duration of soil compaction impact (yr). A_{SC} is calculated as the area thinned and harvested divided by the total area multiplied by the percentage of the surface that experiences soil compaction.

Greenhouse gas emission (CO_2 only)

Greenhouse gases are emitted by two sources: the harvest itself with its combustion of fossil fuel and the natural decomposition of wood that is not harvested. Although the latter is not directly caused by human consumption, it is taken into account here rather than introducing an additional subsystem for natural consumption. [An alternative would be to account only for the net production, i.e. to set $\Delta V_T = 0$ in equation 25 for the no harvest scenario if the production and decomposition of wood are equal].

$$gge_T(H) = (C_V * \Delta V_{NHT}) + (C_F * fu_V * \Delta V_{HT}) \quad [30]$$

where $gge_T(H)$ = greenhouse gas emission due to the removal and decomposition of biomass per unit of time ($kg*yr^{-1}$), ΔV_{NHT} = volume of wood not removed by harvest, but left to decomposition per unit of time (m^3*yr^{-1}), C_F = specific greenhouse gas content of fossil fuel ($kg*kg^{-1}$), fu_V = fuel use for each unit wood removed by harvest ($kg*m^{-3}$), ΔV_{HT} = volume removed by harvest per unit of time.

Financial losses

The financial losses associated with the human consumption subsystem stem from decomposition of biomass, direct cost of harvest and indirect cost of harvest for accidents.

$$fl_T(C) = (Rev_V * \Delta V_{NHT}) + \Delta V_{HT} * prod_{HS}^{-1} (Cost_{PST} + (f_{AHS} * Cost_{Acc})) \quad [31]$$

with $fl_T(C)$ = harvest- and decomposition-induced financial losses ($sFr*yr^{-1}$), Rev_V = average revenue per unit of volume ($sFr*m^{-3}$), $Cost_{PST}$ = cost per system operation hour ($sFr*h^{-1}$), $Cost_{Acc}$ = healthcare cost per accident (sFr).

As mentioned above (cf. 5.1.3), the production results in the generation of financial gains that are capitalized when the products are harvested:

$$fg_T(C) = -(Rev_V * \Delta V_{HT}) \quad [32]$$

with $fg_T(C)$ = growth-induced financial gains capitalized during the harvest per unit of time ($sFR*yr^{-1}$).

In the case of harvest after catastrophe with subsequent storage of wood, equation 32 needs to be adapted as follows:

$$fg_T(C)' = -((Rev_V - Cost_{storV} * \Delta T_{stor}) * us_W * \Delta V_T) \quad [32a]$$

where $Cost_{storV}$ = average storage cost ($sFr*m^{-3}*yr^{-1}$), ΔT_{stor} = storage time, us_W = fraction of usable wood (%).

Transportation Subsystem

Loss of human life

$$lhl_T(Tr) = \Delta V_T * \rho_W * pl^{-1} * d_{TR} * 2 * f_{AccTr} * fr_{Tr} \quad [33]$$

where $lhl_T(Tr)$ = transportation-induced loss of human life per unit of time ($lifes*yr^{-1}$), ρ_W = average density of wood (kgm^{-3}), pl = average payload (kg), d_{TR} = transport distance (km), f_{AccTr} = transportation accident rate (km^{-1}), fr_{Tr} = transportation accident fatality rate (%).

This equation assumes that in every case, the double transport distance needs to be accounted for, i.e. that the transport vehicle travels one way empty every time. While this is a pessimistic assumption, it may not be too far away from real conditions for wood products that must be collected away from places that offer good opportunities for transporting other goods.

For the scenario after a catastrophe, only the usable fraction of the wood is transported.

Disturbance of human health

$$dhh_T(Tr) = \Delta V_T * \rho_W * pl^{-1} * d_{TR} * 2 * f_{AccTr} * ir_{Tr} * d_{AccTr} \quad [34]$$

where $dhh_T(Tr)$ = transport-induced disturbance of human health per unit of time (person*days*yr^{-1}), ir_{Tr} = rate of accidents with seriously injured people needing hospital treatment, d_{AccTr} = average duration of disablement due to transportation accidents (d).

Greenhouse gas emission (CO_2 only)

$$gge_T(Tr) = \Delta V_T * \rho_W * d_{TR} * 1.5 * C_F * fu_{Tr} \quad [35]$$

where $gge_T(Tr)$ = greenhouse gas emissions per unit of time due to the transportation (kg*yr^{-1}), fu_{Tr} = fuel use for transportation (kg*t^{-1}*km^{-1}). The factor 1.5 is to account for the return of the empty vehicle.

Financial losses

Transportation causes direct financial losses in the form of system cost and indirect losses that are due to accidents (repair and healthcare):

$$fl_T(Tr) = \Delta V_T * \rho_W * pl^{-1} * d_{TR} * 2 * (Cost_{Tr} + (f_{AccTr} * (Cost_{Rep} + (ir_{Tr} * Cost_{HC})))) \quad [36]$$

with = financial losses due to transportation per unit of time (sFr*yr^{-1}), $Cost_{Tr}$ = direct cost (sFr*km^{-1}), $Cost_{Rep}$ = average repair cost per accident (sFr), $Cost_{HC}$ = average healthcare cost per accident with injured people (sFr).

Risks associated with substituting production

Substituting the production does not add new subsystems or activities as sources of risk. Thus, the same calculations as outlined in section 5.1 to 5.3, and the risks associated with the substitute production, consumption and transportation systems can simply be added to the ones associated with the respective non-substituted systems:

$$R_{tot}(var\ i, act\ j) = R_{nsp}(var\ i, act\ j) + R_{sp}(var\ i, act\ j)\ \forall\ var, act \quad [37]$$

with R_{tot}(var i, act j) = total risk with regard to a certain risk variable i and activity j, R_{nsp}(var i, act j) = risk associated with the non-substituted production, R_{sp}(var i, act j) = risk associated with the substituted production.

Results

Risks on the Forest Enterprise Level

On the level of the forest enterprise, only the two scenarios harvest and no harvest need to be compared, i.e. risks associated with substituting the production elsewhere are not taken into account.

Tab. 5 and 6 present the results of the analysis for the two scenarios and the two operational states. The variables lsf(G), lsf(C) and lsf(R) are always zero and therefore omitted from the result tables.

Normal operation			
Variable	harvest	no harvest	unit
gget(G)	-9.25E+05	-9.25E+05	kg/yr
lhlt(H)	2.00E-03	2.00E-05	fatalities/yr
dhht(H)	1.00E+00	1.00E-02	persondays/yr
lef(H)	6.00E-01	0.00E+00	ha
lsf(H)	3.30E-01	0.00E+00	%
gget(H)	7.86E+03	9.25E+05	kg/yr
flt(C)	6.10E+04	8.54E+04	sFr/yr
fgt(C)	-8.50E+04	0.00E+00	sFr/yr
lhlt(Tr)	7.50E-05	0.00E+00	fatalities/yr
dhht(Tr)	3.75E-03	0.00E+00	persondays/yr
gget(Tr)	1.18E+04	0.00E+00	kg/yr
flt(Tr)	2.01E+04	0.00E+00	sFr/yr
Loss of human life	*2.08E-03*	*2.00E-05*	*fatalities/yr*
Disturbance of human health	*1.00E+00*	*1.00E-02*	*persondays/yr*
Loss of ecological functionality	*6.00E-01*	*0.00E+00*	*ha*
Loss of soil fertility	*3.30E-01*	*0.00E+00*	*%*
Emission of greenhouse gases	*-9.05E+05*	*2.20E+01*	*kg/yr*[1]
Financial losses	*-3.93E+03*	*8.54E+04*	*sFr/yr*

Table 5. Analysis of the risks associated with the harvest and the no harvest scenario on the forest enterprise level for normal operation.

1 This minimal difference between gget(G) and gget(H) for the no harvest scenario does not show up in the rounded raw results. It is due to the maintenance cutting that is necessary (for opening trails etc.).

There are large relative, but small absolute differences between the harvest and no harvest scenario with regard to the loss of human life and disturbance of human health factor. The difference between the two scenarios is more pronounced when it comes to the loss of soil fertility and ecological functionality. Four these four factors, the no harvest scenario has a risk advantage (lower risk numbers). However, the harvest scenario scores much better with respect to the greenhouse gas emissions and the financial losses, where a negative risk (i.e. a benefit or safety improvement) results as opposed to the positive numbers for the no harvest scenario. This advantage of harvesting is due to the removal of bound CO_2 (gget(H))and the generation of marketable goods (fgt(C)) that results from wood extraction.

	Operation after catastrophe		
Variable	harvest	no harvest	unit
gget(G)	-9.25E+05	-9.25E+05	kg/yr
lhlt(H)	3.00E-02	6.00E-05	fatalities/yr
dhht(H)	7.50E+00	1.50E-01	persondays/yr
lef(H)	2.00E+00	0.00E+00	ha
lsf(H)	5.00E-01	0.00E+00	%
gget(H)	1.89E+06	4.63E+06	kg/yr
flt(C)	4.78E+05	4.29E+05	sFr/yr
fgt(C)'	-1.97E+05	0.00E+00	sFr/yr
lhlt(Tr)	2.25E-04	0.00E+00	fatalities/yr
dhht(Tr)	1.13E-02	0.00E+00	persondays/yr
gget(Tr)	3.53E+04	0.00E+00	kg/yr
flt(Tr)	6.02E+04	0.00E+00	sFr/yr
Loss of human life	*3.02E-02*	*6.00E-05*	*fatalities/yr*
Disturbance of human health	*7.51E+00*	*1.50E-01*	*persondays/yr*
Loss of ecological functionality	*2.00E+00*	*0.00E+00*	*ha*
Loss of soil fertility	*5.00E-01*	*0.00E+00*	*%*
Emission of greenhouse gases	*1.00E+06*	*3.70E+06*	*kg/yr*
Financial losses	*3.41E+05*	*4.29E+05*	*sFr/yr*

Table 6. Analysis of the risks associated with the harvest and the no harvest scenario on the forest enterprise level for operation after catastrophe.

Again, the no harvest scenario has a small to moderate advantage for the loss of human life, disturbance of human health, loss of ecological functionality and loss of soil fertility variables, but a large disadvantage for the greenhouse gas emission and financial loss variables. As a difference to the normal operation, there are no more negative risks for either scenario because of the large amount of wood that is left to decomposition and the depreciation of the wood revenues and the increase in harvesting cost.

Risks on the National Level

On the national level, besides the harvest and no harvest scenario, the possible substitution from within Switzerland is also of interest. Tables 7 and 8 show the risks associated with the this third possibility.

Tab. 11 at the end of the results section shows the aggregation of the risks using the trade-off relations shown in Table 4.

Normal operation Variable	*Substitute system*	*Original system*	***Total***	unit
gget(G)	-9.25E+05	-9.25E+05	-1,85E+06	kg/yr
lhlt(H)	2.00E-03	2.00E-05	2,02E-03	fatalities/yr
dhht(H)	1.00E+00	1.00E-02	1,01E+00	persondays/yr
lef(H)	6.00E-01	0.00E+00	6,00E-01	ha
lsf(H)	3.30E-01	0.00E+00	3,30E-01	%
gget(H)	7.86E+03	9.25E+05	9,33E+05	kg/yr
flt(C)	6.10E+04	8.54E+04	1,46E+05	sFr/yr
fgt(C)	-8.50E+04	0.00E+00	-8,50E+04	sFr/yr
lhlt(Tr)	1.13E-04	0.00E+00	1,13E-04	fatalities/yr
dhht(Tr)	5.63E-03	0.00E+00	5,63E-03	persondays/yr
gget(Tr)	1.77E+04	0.00E+00	1,77E+04	kg/yr
flt(Tr)	3.01E+04	0.00E+00	3,01E+04	sFr/yr
Loss of human life	*2.11E-03*	*2.00E-05*	*2,13E-03*	*fatalities/yr*
Disturbance of human health	*1.01E+00*	*1.00E-02*	*1,02E+00*	*persondays/yr*
Loss of ecological functionality	*6.00E-01*	*0.00E+00*	*6,00E-01*	*ha*
Loss of soil fertility	*3.30E-01*	*0.00E+00*	*3,30E-01*	*%*
Emission of greenhouse gases	*-8.99E+05*	*2.20E+01*	*-8,99E+05*	*kg/yr*
Financial losses	*6.10E+03*	*8.54E+04*	*9,15E+04*	*sFr/yr*

Table 7. Analysis of the risks associated with the harvest and the no harvest scenario on the national level for normal operation.

The substitute system has the same risk profile as the original system for all subsytems except the transportation subsystem [Tr]. Due to the increased transport distance, the associated risks also increase. For the risk analysis on the national level, the risks of the original system (which is now subject to a no harvest treatment) must be added. Since they are all non-negative, the substitution results either in the risk being constant or increasing for all risk variable totals.

Operation after catastrophe Variable	*Substitute system*	*Original system*	***Total***	unit
gget(G)	-2.78E+06	-9.25E+05	-3,70E+06	kg/yr
lhlt(H)	6.00E-03	6.00E-05	6,06E-03	fatalities/yr
dhht(H)	3.00E+00	1.50E-01	3,15E+00	persondays/yr
lef(H)	1.80E+00	0.00E+00	1,80E+00	ha
lsf(H)	9.90E-01	0.00E+00	9,90E-01	%
gget(H)	2.36E+04	4.63E+06	4,65E+06	kg/yr
flt(C)	1.83E+05	4.29E+05	6,12E+05	sFr/yr
fgt(C) / fgt(C)'	-2.55E+05	0.00E+00	-2,55E+05	sFr/yr
lhlt(Tr)	3.38E-04	0.00E+00	3,38E-04	fatalities/yr
dhht(Tr)	1.69E-02	0.00E+00	1,69E-02	persondays/yr
gget(Tr)	5.30E+04	0.00E+00	5,30E+04	kg/yr
flt(Tr)	9.03E+04	0.00E+00	9,03E+04	sFr/yr
Loss of human life	*6.34E-03*	*6.00E-05*	*6,40E-03*	*fatalities/yr*
Disturbance of human health	*3.02E+00*	*1.50E-01*	*3,17E+00*	*persondays/yr*
Loss of ecological functionality	*1.80E+00*	*0.00E+00*	*1,80E+00*	*ha*
Loss of soil fertility	*3.30E-01*	*0.00E+00*	*3,30E-01*	*%*
Emission of greenhouse gases	*-2.70E+06*	*3.70E+06*	*1,00E+06*	*kg/yr*
Financial losses	*1.83E+04*	*4.29E+05*	*4,47E+05*	*sFr/yr*

Table 8. Analysis of the risks associated with the harvest and the no harvest scenario on the national level for operation after catastrophe.

After a catastrophe, the substitution of the production is an advantage with regard to the variables loss of human life, disturbance of human health, loss of ecological and loss of soil fertility. This is due to the reduced probability of accidents and to the more organized way of harvesting in the substitute system (that is not directly affected by the catastrophe). The greenhouse gas emissions change only marginally with the substitution (not observable in the rounded results). However, with regard to the financial losses, the substitution is a disadvantage because of the loss of revenues for the wood that is left to decomposition.

Risks on the Global Level

On the global level, all the possible substitution variants are of interest, while the original no harvest scenario without substitution is no longer relevant if it is assumed that the demand for and the consumption of wood products is not affected. Tables 9 and 10 compare the risks associated with the six possible places of production. Table 11 shows the results of the aggregation using the trade-offs from Table 4 for all three analysis levels.

Normal operation Variable	*Substitute adj. countr.*	*Substitute boreal high*	*Substitute boreal low*	*Substitute tropical*	unit
gget(G)	-1.85E+06	-1.85E+06	-1.85E+06	-1.85E+06	kg/yr
lhlt(H)	2.02E-03	2.60E-04	2.02E-03	1.20E-02	fatalities/yr
dhht(H)	1.01E+00	2.50E-01	1.01E+00	3.01E+00	persondays/yr
lef(H)	6.00E-01	2.40E+00	4.00E+00	3.00E+00	ha
lsf(H)	3.30E-01	5.50E-01	7.50E-01	9.00E-02	%
gget(H)	9.33E+05	9.33E+05	9.33E+05	9.33E+05	kg/yr
flt(C)	1.31E+05	1.26E+05	1.16E+05	1.48E+05	sFr/yr
fgt(C)	-8.50E+04	-8.50E+04	-8.50E+04	-8.50E+04	sFr/yr
lhlt(Tr)	3.60E-04	3.60E-04	3.60E-04	4.50E-03	fatalities/yr
dhht(Tr)	1.80E-02	1.80E-02	1.80E-02	2.25E-01	persondays/yr
gget(Tr)	4.71E+04	2.36E+05	2.36E+05	4.52E+05	kg/yr
flt(Tr)	7.23E+04	1.02E+05	9.41E+04	6.41E+04	sFr/yr
Loss of human life	*2.38E-03*	*6.20E-04*	*2.38E-03*	*1.65E-02*	*fatalities/yr*
Disturbance of human health	*1.03E+00*	*2.68E-01*	*1.03E+00*	*3.24E+00*	*persondays/yr*
Loss of ecological functionality	*6.00E-01*	*2.40E+00*	*4.00E+00*	*3.00E+00*	*ha*
Loss of soil fertility	*3.30E-01*	*5.50E-01*	*7.50E-01*	*9.00E-02*	*%*
Emission of greenhouse gases	*-8.70E+05*	*-6.81E+05*	*-6.81E+05*	*-4.65E+05*	*kg/yr*
Financial losses	*1.19E+05*	*1.43E+05*	*1.26E+05*	*1.28E+05*	*sFr/yr*

Table 9. Analysis of the risks associated with the harvest and the no harvest scenario on the global level for normal operation.

With regard to the variables loss of human life and disturbance of human health, substitution is an advantage in the case of highly mechanized boreal production, but a disadvantage in the case of tropical production when compared to the original system (cf. Table 5). All boreal and tropical substitution solutions are clearly a disadvantage with regard to the loss of ecological functionality variable, but the tropical substitution means a smaller loss of soil fertility. Compared to the original system, all substitutions systems increase the greenhouse gas emissions, and the same also applies to the financial losses.

Operation after catastrophe Variable	*substitute adj. countr.*	*substitute boreal high*	*substitute boreal low*	*substitute tropical*	unit
gget(G)	-3.70E+06	-3.70E+06	-3.70E+06	-3.70E+06	kg/yr
lhlt(H)	6.06E-03	7.80E-04	6.06E-03	3.61E-02	fatalities/yr
dhht(H)	3.15E+00	8.70E-01	3.15E+00	9.15E+00	persondays/yr
lef(H)	1.80E+00	7.20E+00	1.20E+01	9.00E+00	ha
lsf(H)	3.30E-01	5.50E-01	7.50E-01	9.00E-02	%
gget(H)	4.65E+06	4.65E+06	4.65E+06	4.65E+06	kg/yr
flt(C)	5.67E+05	5.50E+05	5.22E+05	6.18E+05	sFr/yr
fgt(C) / fgt(C)'	-2.55E+05	-2.55E+05	-2.55E+05	-2.55E+05	sFr/yr
lhlt(Tr)	1.08E-03	1.08E-03	1.08E-03	1.35E-02	fatalities/yr
dhht(Tr)	5.40E-02	5.40E-02	5.40E-02	6.75E-01	persondays/yr
gget(Tr)	1.41E+05	7.07E+05	7.07E+05	1.35E+06	kg/yr
flt(Tr)	2.17E+05	3.06E+05	2.82E+05	1.92E+05	sFr/yr
Loss of human life	*7.14E-03*	*1.86E-03*	*7.14E-03*	*4.96E-02*	*fatalities/yr*
Disturbance of human health	*3.20E+00*	*9.24E-01*	*3.20E+00*	*9.83E+00*	*persondays/yr*
Loss of ecological functionality	*1.80E+00*	*7.20E+00*	*1.20E+01*	*9.00E+00*	*ha*
Loss of soil fertility	*3.30E-01*	*5.50E-01*	*7.50E-01*	*9.00E-02*	*%*
Emission of greenhouse gases	*1.09E+06*	*1.66E+06*	*1.66E+06*	*2.30E+06*	*kg/yr*
Financial losses	*5.29E+05*	*6.01E-05*	*5.50E+05*	*5.56E+05*	*sFr/yr*

Table 10. Analysis of the risks associated with the harvest and the no harvest scenario on the global level for operation after catastrophe.

After a catastrophe, substitution decreases the risks of loosing human lives and disturbing human health except for the case of tropical substitution, where an increase results compared to the original system (cf. Table 6). However, substitution is also associated with an increased loss of ecological functionality for the boreal and tropical cases, but a slight decrease for the two other cases. The loss of soil fertility is particularly pronounced if the substitute production is from boreal regions, but it decreases in the case of tropical substitution. The emission of greenhouse gases increases in all cases of substitution, although only slightly if the increase in transport distances is moderate. The financial losses also increase for all substitutes.

Scenario Operational state	*Harvest*	*No harvest, no subtitute*	*Substitute CH*	*Substitute adj. ctr.*	*Substitute boreal high*	*Substitute boreal low*	*Substitute tropical*
Normal operation	-0.0947	0.0157	-0.0756	-0.0637	0.1249	0.2784	0.2216
Operation after catastrophe	0.6605	0.9030	0.5401	0.5763	1.0844	1.4914	1.4954

Table 11. Aggregated risk (expressed as risk equivalent value REV, cf. Eq. 21) of the harvest and no harvest scenario without and with substitution for normal operation and operation after a catastrophe.

The aggregation shows that the harvest scenario has the lowest associated REV for normal operation. In cases where the place of production (source) is close to the place of processing (sink), even a negative risk (benefit or safety improvement) can be achieved. For the operation after catastrophe, substituting the production in nearby areas can reduce the risk slightly, but a negative risk is no longer achievable.

Discussion and Conclusions

General Findings

The sample case study presented shows that the framework and the methodology of a risk analysis can in fact be used to evaluate forest policy and management options. The risk analysis provides a transparent procedure that is easy to implement on different levels and for a variety of geographical settings. It is reproducible, and it can therefore be used as a tool for supporting the decision-making process (where risk is usually but one of several factors to be considered) and as a basis for discussion. Thus, the present study may also be understood as a proof of feasibility.

One of the main advantages of the risk analysis approach for policy evaluation is its flexibility in handling a variety of input information and in generating the desired output risk variables. This is particularly useful for complex systems, where empirical data is often missing and has to be replaced by expert judgment or by assumptions.

The risk analysis methodology is applied to a well-defined system. Everything that is not explicitly part of the analysis system is by definition not considered, and it is therefore less probable that external effects are omitted unintentionally. A risk analysis also depends on explicit formulation of qualitative and quantitative relations, i.e. all the data used in the analysis need to be disclosed. In some instances, it is difficult to do this, e.g. in the case of fuzzy variables such as the ecological functionality used herein. However, in most cases the operationalization of preferences and values, which is a byproduct of this formulation, outweighs the problem by far. Discussions can focus on the system components and relations used within the analysis rather than on subjective opinions about individual and societal values and preferences.

One potential problem of using risk analysis for policy evaluation is the form of the results. The fact that the risk is eventually represented as a single number may on the one hand suggest an accuracy that cannot be achieved in most cases; on the other hand, accepting the result requires a certain "trust in numbers" that many people affected by the policy decisions do not have. This may at least partly be due to the fact that the numbers do not reveal the values and preferences that are at the basis of the assumptions made for the calculation. However, even if the acceptance of the risk analysis results is sometimes questionable, they still provide a solid and reproducible foundation for a discussion about risks.

Method

The sample case has shown that the definition and operational formulation of risk variables can be critical. Fuzzy or "soft" variables must be expressed on an absolute or a relative scale and measurement units must be assigned to reference them in space and time. This is usually not a problem for a single variable, but nontrivial and non-obvious trade-off relations may be required as soon as an aggregation is performed.

Another problem is the mathematical characteristics of the risk variables. In the sample case, linear expressions have been used for all variables, and it is questionable whether this is in compliance with the risk perception and evaluation by the public. Logarithmic expressions are probably more suitable for representing a risk whose probability of occurrence and whose consequences extend over several orders of magnitude. However, the risks associated with the different scenarios investigated herein are all very small because their consequences do not reach catastrophic dimensions, and within this scape, a linear representation is sufficiently accurate.

Comparing systems on different levels as done in the study requires that the same model representation is used for all of them to ensure consistency. Considering the different amount and quality of data that is available, it becomes soon difficult to provide models that are simple enough to work on all levels, yet qualitatively acceptable. On the other hand, if more sophisticated models are used, then the necessary information may no longer be available for all levels of investigation because of the different resolutions of the systems under consideration.

Data

The data requirements follow more or less directly from the definition of the systems and the formulation of models for the calculation and they are not difficult to derive. However, acquiring the information is the biggest problem in the present study. This is not unusual for the risk analysis of complex technical or environmental systems, but surprising for such a simple system as the wood production in a forest enterprise in Switzerland. In some cases, this can limit the applicability of risk analysis for policy evaluation, because people may criticize a methodology that is based on so many assumptions as shown in the example. However, it can also be considered a merit of the risk analysis methodology to reveal all the aspects where knowledge about the risky processes that occur in a system is lacking. More traditional methods of policy evaluation do not necessarily use fewer assumptions, they are often just not openly declared.

In practice, it is usually possible to acquire more and better information for a concrete analysis than shown in the example, but often at a higher cost. For example, data are often available in a statistically more meaningful form than presented here, i.e. as distributions. Unfortunately, the cost of acquiring the data rises generally faster than the quality of the information, and providing a complete and detailed knowledge base is often just not affordable. This is probably one of the most important factors in limiting the applicability of risk analysis for policy evaluation in forestry.

For the aggregation step, the sources for comparing risks are relatively scarce, and for most of the applications, common practice and the few legal documents available are the only basis for developing an aggregation scheme. Even if existing documents give an indication how risks expressed in different variables might be compared, they can rarely be applied as-is for the purpose at hand, and an adaptation and / or transformation is required. It is obvious that subjective preferences have an influence on this process, and the aggregation scheme can never be considered an "objective" method.

Calculations

The formulations of the equations and the calculations themselves do not pose problems, since most of the equations are simple multiplications and since only point estimates are available for most of the parameters. In cases like the example shown here, the computation can even be performed manually in a spreadsheet model, thus potentially allowing for sensitivity calculations and modeling of stochastic input data.

One problem arises when the risks for the operation after a catastrophic event are calculated. The amount of salvage wood (5000 m^3) is a multiple of the annual production (1000 m^3) that is the basis for the sustainability considerations. In other words, harvesting the salvage wood corresponds to a five-year period of sustainable normal operation. It is not clear how this can be accounted for in the calculations, e.g. for the risk variables that refer to a certain observation period. In the calculations for the harvest scenario, it was assumed that all the salvage wood is harvested within the first year and that the usable part is removed. To compensate for this over-exploitation, four subsequent years without any harvest activity would be required, i.e. the observation period could be extended to five years. However, for the risk calculation, using one year as a basis rather errs on the safe (i.e. pessimistic) side, and this is preferable for the comparison.

Results

The results show three interesting facts. First, for the normal operation, it is less risky to harvest wood than not to harvest it or to substitute the production elsewhere if the assumptions in Tables 2 and 3 are used for the calculation. This is primarily due to the greenhouse gas emission factor and secondarily to the loss of ecological function factor, i.e. one or two components have a critical influence on the overall result. This shows the importance of carefully selecting an aggregation scheme. It can therefore be argued that the trade-off ($4,5*10^6$ kg CO_2 risk equivalent to 1 human life) is not realistic and does distort the calculation. However, it is not uncommon in a risk analysis that a single component dominates the whole risk profile, and there is no immediate need to change the trade-offs.

Second, the risk depends strongly on the system under consideration. The difference in risk between the two scenarios harvest and no harvest is not the same for the enterprise level (without substitution) as for the national level (with substitution). Not taking this difference into account leads to distorted and wrong conclusions. It is thus very important for a risk analysis to properly define what systems and effects are covered. This also has an implication for the policy selecting process. It must be ensured that the consequences of selecting a certain policy are assessed not only on the level on which the policy-setting entity is competent, but also on superior levels. If this is not the case, then the risks associated with an activity (as producing wood) are perhaps simply transferred to another place where the preference set differs, most often as a consequence of economic pressures. In other words, something that is considered unacceptably risky e.g. in Switzerland is then done perhaps in an even more risky way in countries like Russia or Malaysia, where aspects such as the protection of the environment or occupational safety and health are considered less important than revenues from wood sales. A scale effect can naturally also result if the analysis is transferred to lower levels (e.g., on the forest division or stand level). This could yield valuable information for making land-use decisions within an enterprise, i.e. for allocating the land base to different, mutually incompatible uses.

Third, the operational state considered has an influence on the policy making. After a catastrophe, it may be less risky to substitute the production elsewhere than to harvest the salvage wood. This is due to the increased accident and fatality rates and to increased greenhouse gas emissions. It can now certainly not be implied that salvage wood after a catastrophe should generally not be harvested (in many cases, a harvest is a technological condition for future forestry activities such as planting), but that exceptional situations require consideration in the decision-making process. The feasibility and other relevant constraints need to be considered when defining the policy options that are to be evaluated.

Summarizing, in the sample case study, the risk analysis methodology did clearly reveal some critical points that would probably be much more difficult to identify if a classical policy evaluation procedure was applied. Its applicability to a wide variety of systems and investigation levels combined with its transparency and accuracy make it a powerful tool for analyzing forest management and policy options for the decision-maker. However, risk analysis cannot replace the political assessment on a scope beyond the control of the decision-making entity.

References

Beroggi, G.; Kröger, W., 1993: Risikoanalyse technischer Systeme: Methoden, Modelle, Verfahren und Hilfsmittel. SIA Schweizer Ingenieur und Architekt 47: 877-883.

BFS 1998: Strassenverkehrsunfälle in der Schweiz 1996. Statistik der Schweiz Fachbereich 11: Verkehr und Nachrichtenwesen. BFS, Bern.

Briggs, D. J. & Wills, J., 1999: Presenting Decision-Makers with their Choices: Environment Health Indicators for NEHAPs. In: Briggs, D. J., Stern, R., Tinker, T. L.: Environmental Health for All.

Briggs, D. J., Stern, R. & Tinker, T.L., 1999: Environmental Health for All. Risk Assessment and Risk Communication for National Environmental Health Action Plans. NATO Science Series Environmental Security Vol. 49. Kluwer Academic Publishers, Dordrecht.

Brouwer, L. C., 1996: Nutrient Cycling in Pristine and Logged Tropical Rain Forest. A Study in Guyana. Thesis Universiteit Utrecht. Tropenbos Guyana Series 1. Elinkwijk, Utrecht.

Bunn, D. W., 1984: Applied Decision Analysis. McGraw-Hill, New York.

Butora, V., 1978: Das Unfallgeschehen in der Forstwirtschaft. Eine Auswertung der Arbeitsunfälle in den öffentlichen Waldungen der Schweiz. Institut für Hygiene und Arbeitsphysiologie, ETH Zürich. Forstwirtschaftliche Zentralstelle der Schweiz, Solothurn.

BUWAL, 1991, 1992: Handbuch I / II / III zur Störfallverordnung, StFV. EDMZ, Bern.

BUWAL, 1998: Methoden zur Analyse und Bewertung von Naturgefahren. Eine risikoorientierte Betrachtungsweise. EDMZ, Bern.

BUWAL, 1999a: Risikoanalyse bei gravitativen Naturgefahren. Bd. 1: Methode. EDMZ, Bern.

BUWAL, 1999b: Risikoanalyse bei gravitativen Naturgefahren. Bd. 2: Fallbeispiele und Daten. EDMZ, Bern.

Culmann, C., 1864: Bericht an den hohen schweizerischen Bundesrath über die Untersuchung der schweiz. Wildbäche, vorgenommen in den Jahren 1858, 1859, 1860 und 1863. Zürcher und Furrer, Zürich.

FIO, 1993: Viviandenken: Vivian stellte alle Beteiligten vor neue Probleme, die auch ein neues Denken erforderten – Vivian-Andenken: Ein Andenken an Vivian, den Jahrhundertsturm von 1990 und seine Bewältigung. Forstinspektion Oberland, Spiez..

Froehlich, H. A., Miles, D. W. R. & Robbins, R. W., 1986: Growth of young Pinus ponderosa and Pinus contorta on compacted soil in central Washington. Forest Ecology and Management 15: 285 – 294.

Hennen, L., 1995: Technikfolgen-Abschätzung als Instrument der Umweltpolitik. In: Joussen, W., Hessler, A. A.: Umwelt und Gesellschaft. Eine Einführung in die sozialwissenschaftliche Umweltforschung. Akademie Verlag , Berlin.

Hessler, A.A., 1995: Angst um die Umwelt? Ein Streifzug durch die Risikopolitik. In: Joussen, W., Hessler, A. A.: Umwelt und Gesellschaft. Eine Einführung in die sozialwissenschaftliche Umweltforschung. Akademie Verlag , Berlin.

Hofstetter, P., 1998: Perspectives in Life Cycle Impact Assessment. Kluwer Academic Publishers, Boston, MA.

Hollenstein, K., 1997: Analyse, Bewertung und Management von Naturrisiken. Vdf Hochschulverlag Zürich.

Jonkers, W. B. J., 1987: Vegetation Structure, Logging Damage and Silviculture in a Tropical Rain Forest in Suriname. Agricultural University Wageningen.

Linstone, H. A., 1984: Multiple Perspectives for Decision Making. Elsevier Science Publishers, New York.

Löffier, H. & Ohmer, G., 1994: Steigerung der Arbeitssicherheit im Kleinprivatwald "Odenwaldprojekt". Schriftenreihe des Bundesministeriums für Ernährung, Landwirtschaft und Forsten Serie A: 437. Landwirtschaftsverlag, Münster.

Lootsma, F. A., 1999: Multi-Criteria Decision Analysis Via Ratio and Difference Judgement. Kluwer Academic Publishers, Dordrecht.

Meier, M., 1992:"Man gewöhne sich mit der Zeit daran" – Risiken im Basel des 19. Jahrhunderts. In: Risiko zwischen Chance und Gefahr. Edited by Locher, R. Reto Locher & Partner, Basel.

Minssen, H., 1878: Statistik der Dampfkessel-Explosionen im Deutschen Reiche 1877. Mittheilungen aus der Praxis des Dampfkessel- und Dampfmaschinenbetriebes 1: 133 – 134.

Morris, L. A. & Miller, R. E., 1994: Evidence for long-term productivity change as provided by field trials. In: Impacts of Forest Harvesting on Long-Term Site Productivity. Edited by Dyck, W. J., Vole, D. W., Comerford, N. B. Chapman and Hall, London etc.

NRC, 1975: Reactor Safety Study, An Assessment of Accident Risks in U.S. Commercial Nuclear Power Plants, U.S. Atomic Energy Commission. Washington, D.C.

Petak, W. J. & Atkisson, A. A., 1982: Natural Hazard Risk Assessment and Public Policy. Springer Series on Environmental Management. Springer, New York.

PSW, 1996: Appendices [to the] Revised Draft Environmental Impact Statement: Managing California Spotted Owl Habitat in the Sierra Nevada National Forests of California – An Ecosystem Approach. USDA Forest Service Pacific Southwest Region.

Rehschuh, D., Gerdsen, G., Hartfiel, J., Mattes, H. & Scheele, H.A., 1984: Unfallursachenforschung zur Erhöhung der Arbeitssicherheit in der Forstwirtschaft. Kuratorium für Waldarbeit und Forsttechnik / Schriftenreihe der Bundesanstalt für Arbeitsschutz: Fb384.

Ruth, L., Wutzke, J., Allen, J., Nielsen, K. & Batts, D., 1994: Conserving the California Spotted Owl: Impacts of Interim Policies and Implications for the Long-Term. UC Davis Wildland Resources Center Report 33.

Saaty, T. L., 1980: The Analytic Hierarchy Process. Planning, Priority Setting, Resource Allocation. McGraw-Hill, New York.

Schweizer Rück, 1991: Historische Erdbeben in Europa. Schweizerische Rückversicherungsgesellschaft, Zürich.

Senyk, J. & Craigdaille, D., 1997: Effects of harvesting methods on soil properties and forest productivity in interior British Productivity. Pacific Forestry Centre Information Report BC-X-365. Victoria, B.C.

SEV, 1893: Statuten des Schweizerischen elektrotechnischen Vereins. Jahrbuch des Schweizerischen elektrotechnischen Vereins 4: 15 – 18.

von Neumann, J. & Morgenstern, O., 1944: Theory of Games and Economic Behavior. Princeton University Press, Princeton NJ.

Weimer, D. L. & Vining, A. R., 1992: Policy Analysis. Concepts and Practice. Prentice Hall, Englewood Cliffs.

Weiss, G., 1997: Evaluierung der Politik zum Schutz vor Naturgefahren – Grundlagen der Evaluation und Forschungsansatz. Centralblatt für das gesamte Forstwesen 114: 189 – 211.

Williamson, J. R., 1990: The effects of mechanised forest harvesting operations on soil properties and site productivity. Tasmanian Forest Research Council Inc. Research Report No 5.

RISK ANALYSIS IN THE CONTEXT OF TIMBER HARVEST PLANNING

Atsushi Yoshimoto[1]

ABSTRACT

Events related to forest management are associated with uncertainty. This affects the management decision for forest level planning as well as stand level planning. In this paper, two stochastic models are introduced. The first model is a macro-type stochastic model for timber supply analysis by predicting harvesting events over the time horizon, while the second model is a micro-type stochastic model used to search for an optimal rotation for forest stand management under price uncertainty. A macro-type approach describes private forest owners' harvesting behavior with the use of a stochastic process. The threshold method was assumed for the decision-making mechanism. In contrast, a micro-type approach was to search for an optimal rotation period for forest stand management under price uncertainty. Unlike the macro type approach, uncertainty was assumed for log price, which affects an optimal decision for the stand management. Stochastic differential equations are applied to capture stochasticity of price dynamics, and a stochastic dynamic programming approach is utilized for optimization. The continuous time stochastic process, i.e. the geometric Brownian motion, is used to model the log price process. The binomial option pricing approximation is then applied to discretize the price dynamics. Estimation of nonlinear models to stochastic phenomena is also elaborated.

Keywords: *Forest economics, uncertainty, stochastic model, stochastic differential equation, stochastic dynamic programming*

[1] Dept. of Statistical Methodology, Institute of Statistical Mathematics, Tokyo, 106-8569, JAPAN, e-mail: yoshimoa@ism.ac.jp

K. von Gadow (ed.), Risk Analysis in Forest Management, 113–148.

INTRODUCTION

Events related forest management are associated with uncertainty. Uncertainty exists in log price and various cost observations as well as forest growth affected by environmental variations. This uncertain management environment affects timber supply as well as forest stand management. The objective of this paper is to present stochastic models to deal with timber supply analysis as well as forest stand management.

Forecasting timber supply is a major issue in forest economics. Depending upon the timber supply, the price of logs, lumber, and other forest products varies. In North America, examples of timber supply models include TAMM (*T*imber *A*ssessment *M*arket *M*odel) by Adams and Haynes (1980) and PAPYRUS by Gilless and Buongiorno (1987). These models are based on mathematical programming techniques such as linear programming and quadratic programming as well as econometric models of timber demand and supply. Because of the nature of the techniques applied, all parameters and variables in the above models are deterministic, and there is no consideration of uncertainty into the model.

A stochastic modeling approach can be applied for timber supply analysis. Considering forest owners' harvesting events as a stochastic event, it becomes natural to incorporate stochastic factors into the timber supply model. In Japan back in the 1960's, the Science and Technology Agency (1961, 1963) introduced a stochastic timber supply model. This was the first stochastic model applied for the timber supply analysis (Suzuki 1972, 1984). The model has been used for predicting not only the nationwide timber supply, but also the regional timber supply in Japan. Applications of this stochastic timber supply model can be found in Amano et al. (1984) and Blandon (1985, 1991, 1993). Although the model incorporates the stochastic nature of timber supply, it has not been flexible enough to reflect change in the forest owner's harvesting behavior affected by change of economic factors, such as price, interest rate, and costs. Since stochastic modeling of harvesting behavior provides the structural insight to investigate when harvesting events will occur or how many such events will occur within a certain period of time, continuous adjustment of the model is necessary.

Assuming the management objective value function for a forest owner's decision, harvesting behavior can be captured by a stochastic process as well. In the next section, the management objective value is assumed to have a nature of stochasticity. Prediction of timber supply is carried out by searching for felling probabilities of forest stands over different age classes, where the threshold method is assumed to control forest owners' harvesting decision. That is, we suppose that a forest owner will fell forest stands whenever the induced management value reaches its threshold.

While the above stochastic model for timber supply regards forest harvesting events as a stochastic event, the next stochastic model for forest stand management deals with a harvest decision making under uncertainty. During the last decades, imported timber to Japan from countries such as the U.S., Canada, and the Nordic countries, has occupied a large share of the domestic timber market. This resulted in a situation where the domestic forest managers have been facing a large degree of future price uncertainty. Because of this, it becomes necessary to take the future price uncertainty into account within the forest management framework.

In application of stochastic modeling under uncertain phenomena, recent development of financial option valuation techniques has been paid a great deal of attention in order to evaluate investment activities under uncertainty. Examples include Brennan and Schwartz (1985) for valuing a copper mining project, Bailey (1991) for valuing agricultural firms. Examples applied to forestry include Zinkhan (1991) for valuing the land-use conversion option from forest lands, and Morck et al. (1989) for valuing a mature forest under a stochastic environment as to price and growth. Other research dealing with stochastic models includes Clarke and Reed (1989), Reed and Clarke (1990), Haight and Holmes (1991), Reed and Ye (1994), Thomson (1992), Yoshimoto and Shoji (1998), Plantinga (1998) and Thorsen (1999). Unlike the macro-type approach, uncertainty is assumed for log price, which affects an optimal decision for the stand management. A stochastic differential equation is applied to capture stochasticity of prices, then a stochastic dynamic programming model is constructed for decision making. We assume that log prices follow the geometric Brownian motion, and that forest stand growth is represented by a deterministic Richards growth function (Richards, 1958).

As can be seen in a wide variety of applications for stochastic modeling for forest stand management, the Brownian motion and geometric Brownian motion assumptions have been used to represent a target stochastic process. There is almost no argument on model selection for a target stochastic process. This might be associated with technical difficulty in estimating other stochastic models. Since any error in stochastic model specification could result in large deviation from a "true" optimum, model specification has to be implemented with care. In the last part of this paper, estimation of nonlinear models to stochastic phenomena is elaborated. Focusing on the log price stochastic behavior, we propose thirteen alternatives for a continuous time stochastic model based on the state dependent volatility process (SDVP), then based on AIC (Akaike, 1977) we investigate if there exists any difference in the performance of each model. We apply the local linearization method to estimate parameters of models.

STOCHASTIC MODEL FOR FORECASTING HARVESTING EVENTS

A stochastic model for timber supply analysis was introduced back in the early 1960's by the Science and Technology Agency (1961, 1963) in Japan. A harvesting event was treated as a random variable which was a function of time, then the probability density function of the events over the time horizon was derived. This probability density function is important because with the density function we can forecast how much timber would be available for harvest at a certain age class by multiplying the total forest areas under management by probability of being harvested at that age class. In this section, we introduce a new stochastic model for forecasting harvesting events incorporating economic factors. The model is based on the idea introduced by the Science and Technology Agency (1961, 1963). Then we present an experimental use of the proposed model.

Timber Supply Projection and Markov Chain

We consider a situation where we have forests classified into different age classes, and where we are only concerned about a harvesting process, i.e., {Harvest} or {Growth}. Let $\underline{X}_t$ be a (n x 1) vector expressing forest stand areas of different age classes at time t, where the i-th element of the vector, x_i^t represents forest stand area at the i-th age class (i = 1, 2,,,n). This means that once we have $\underline{X}_t$, we know the age structure of the forest. For the purpose of simplicity, we assume that only forest stand age governs forest stand growth. In other words, we only consider one-dimensional growth dynamics. Further we introduce a sequence of random variables {y_t, t=1,2,3,,}, representing a state of forest stands or age class, then define $y_t = i$ for the corresponding x_i^t. The possible finite outcomes or age classes of y_t thus becomes j {j=1,2,,,n}. We suppose that whenever a forest stand is in age class i, there is a probability $p_{i,i+1}^t$ that the stand will next be in age class (i+1). That is, we suppose that

$$p_{i,i+1}^t = \Pr\{y_{t+1} = i+1 \mid y_t = i\} \qquad (1)$$

This is the probability that the forest stand grows till age class (i+1) at time (t+1), or transition from the state i at time t to the state (i+1) at time (t+1). The complement of this probability therefore becomes the probability that a forest stand at age class i is harvested in time interval [t,t+1). After harvesting, a forest stand is assumed to be replanted immediately. We denote this state by state 1 or age class 1. Therefore the process of {y_t, t=1,2,3,,} becomes the Markov chain. The diagraph of the chain of a harvesting process considered here is shown in Figure 1.

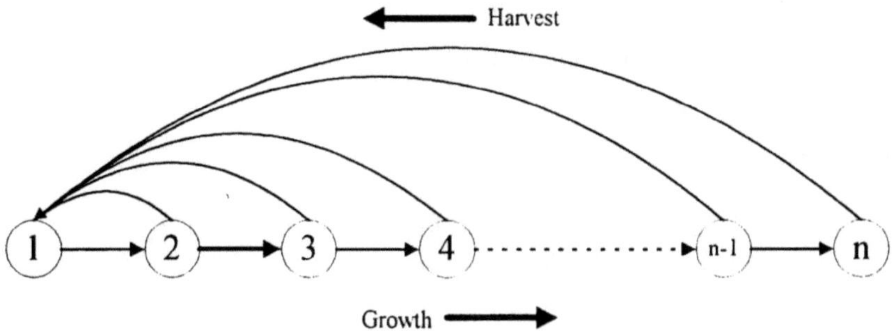

Figure 1. Transition graph of the Markov chain for harvesting process

Introducing the (n x n) square transition probability matrix at time t defined by

$$P_t = \begin{bmatrix} p^t_{1,1} & p^t_{1,2} & 0 & 0 & \ldots & 0 \\ p^t_{2,1} & 0 & p^t_{2,3} & 0 & \ldots & 0 \\ p^t_{3,1} & 0 & 0 & p^t_{3,4} & \ldots & 0 \\ . & . & . & . & \ldots & . \\ 1 & 0 & 0 & 0 & \ldots & 0 \end{bmatrix} \qquad (2)$$

we can predict a forest age structure in terms of area, given an initial age structure, $\underline{X}_0$, by

$$\underline{X}'_{t+1} = \underline{X}'_t \cdot P_t$$

$$\left(x_1^{t+1}\, x_2^{t+1}\, x_3^{t+1} \ldots x_n^{t+1}\right) = \left(x_1^t\, x_2^t\, x_3^t \ldots x_n^t\right) \cdot \begin{bmatrix} p^t_{1,1} & p^t_{1,2} & 0 & 0 & \ldots & 0 \\ p^t_{2,1} & 0 & p^t_{2,3} & 0 & \ldots & 0 \\ p^t_{3,1} & 0 & 0 & p^t_{3,4} & \ldots & 0 \\ . & . & . & . & \ldots & . \\ 1 & 0 & 0 & 0 & \ldots & 0 \end{bmatrix} \qquad (3)$$

where ' represents transpose. Here we assume that all forest stands at age class n are harvested, then converted into state 1 or age class 1 after immediate replantation. If we know forest stand volume per unit area at each age class, we can predict timber supply, V_t, at time t by

$$V_t = \sum_{i=1}^{n} v_i \cdot x_i^t \cdot p_{i,1}^t \qquad (4)$$

where v_i is harvestable volume per unit area at age class i. Here one question arises as to how to estimate all probabilities in the transition matrix.

Counting Process

Let's define the probability, q_i, that a forest stand at age class i will be harvested in the next time period, i.e. time interval $[i, i+1)$ after being planted, and call it the i-th Gentan[2] probability. The probability that the stand at age class i will be harvested in the next time period at time t, i.e. time interval $[t, t+1)$ is represented by the following conditional probability, $p_{i,1}^t$,

$$\begin{aligned} p_{i,1}^t &= \Pr\{y_{t+1} = 1 \mid y_t = i\} \\ &= \frac{\Pr\{y_{t+1} = 1, y_t = i\}}{\Pr\{y_t = i\}} \\ &= \frac{q_i}{1 - \sum_{k=1}^{i-1} q_k} \end{aligned} \tag{5}$$

In order to distinguish from the Gentan probability, we use the term, the i-th felling probability at time t for this conditional probability. With $p_{i,1}^t\ (i = 1, 2,,, n)$, the above transition probability matrix is constructed by a set of the felling probabilities for all age classes and their complements. The felling probability can be interpreted as a percentage of all the stands with stand age i to be harvested in time interval $[t, t+1)$. Up to this point, given a set of the Gentan probabilities for all age classes, we can derive the transition probability matrix, then predict timber supply from forests. In what follows, we introduce a stochastic process which can be used for the Gentan probability estimation.

At first, we apply a counting process to describe the forest owner's decision making mechanism. It is different from the optimization framework, which searches for an optimal decision by either maximizing or minimizing some objective function. In the proposed stochastic model, we utilize the threshold method for the decision making mechanism, where the forest owner will decide to harvest a forest stand when his/her objective value exceeds a given threshold. In order to derive a stochastic process for the forest owner's harvesting behavior, let's assume that a forest owner will wait until the management objective value reaches its threshold, say M, and that an average value function for the objective value at time t is $g(t)$. Here we assume stochasticity for the management objective value.

With use of a concept of counts, let's define one count equal to one unit of increase in the management objective value. We assume that when the objective value increases by one unit, then the forest owner will count one. Letting $N(t)$ be the total number of counts to have occurred by time t, then $N(t)$ can be regarded as the management objective value in a discrete

[2] The term "Gentan" was adapted from the Japanese agricultural policy for rice fields, i.e."reduction in rice fields" in Japanese, and defined by Suzuki (1972)

sense, and is a counting process with the following properties for any time s and t,

i) $N(t) \geq 0$; the total number of counts is non-negative.

ii) $N(t)$ is integer valued; a count is integer, so is the total number of counts.

iii) If $s \leq t$, then $N(s) \leq N(t)$; the total number of counts is non-decreasing over time.

iv) For $s \leq t$, $N(t) - N(s)$ equals the number of counts that have occurred in the interval (s, t).

Here we only assume non-decreasing objective value, and we terminate the process once the objective value decreases.

We further assume the following for a counting process, $N(t)$

v) $N(0) = 0$

vi) $\{N(t), t \geq 0\}$ has independent increments,

vii) $\Pr\{N(t+\Delta t) - N(t) = 1\} = \dfrac{dg(t)}{dt} \cdot \Delta t + o(\Delta t)$

viii) $\Pr\{N(t+\Delta t) - N(t) \geq 2\} = o(\Delta t)$

where $o(\Delta t)$ is used as a symbol to denote a function of Δt, which tends to 0 more rapidly than Δt satisfying,

$$\lim_{\Delta t \to 0} \frac{o(\Delta t)}{\Delta t} = 0 \tag{6}$$

While the assumption v) constraints an initial condition of $N(t)$, the assumption vi) implies that $N(t)$ is independent of the number of harvesting events in an interval prior to the interval [0,t), i.e. future changes in $N(t)$ are independent of the past changes. The assumption vii) can be justified by a consequence of the Poisson approximation to the binomial distribution. The assumption viii) is to show that the probability of more than one occurrence of harvesting events is negligible as Δt approaches zero.

Under these assumptions, we derive a differential-difference equation for the probability that $N(t)$ becomes k, i.e., the number of counts becomes k. To seek the probability density function of $N(t)$, let

$$P_k(t) = \Pr\{N(t) = k\} \tag{7}$$

A differential-difference equation for $P_k(t)$ is derived by considering the probability that the number of counts in time $(t + \Delta t)$ becomes k,

$$\begin{aligned} P_k(t+\Delta t) &= \Pr\{N(t+\Delta t) = k\} \\ &= \Pr\{N(t+\Delta t) - N(t) = 0, N(t) = k\} \\ &\quad + \Pr\{N(t+\Delta t) - N(t) = 1, N(t) = k-1\} \\ &\quad + \sum_{j=2}^{k} \Pr\{N(t+\Delta t) - N(t) = j, N(t) = k-j\} \end{aligned} \tag{8}$$

Since the probability that more than one occurrence is negligible from the assumption viii), the last summation term of the above equation is also $o(\Delta t)$, resulting in,

$$P_k(t+\Delta t) = \{1-\frac{dg(t)}{dt}\Delta t - 2o(\Delta t)\}\cdot P_k(t) + \{\frac{dg(t)}{dt}\Delta t + o(\Delta t)\}\cdot P_{k-1}(t) + o(\Delta t)$$

$$= \{1-\frac{dg(t)}{dt}\Delta t\}\cdot P_k(t) + \frac{dg(t)}{dt}\Delta t\cdot P_{k-1}(t) + o(\Delta t) \tag{9}$$

$$\frac{P_k(t+\Delta t)-P_k(t)}{\Delta t} = -\frac{dg(t)}{dt}\cdot P_k(t) + \frac{dg(t)}{dt}\cdot P_{k-1}(t) + \frac{o(\Delta t)}{\Delta t}$$

Thus, we have the following differential-difference equation for $P_k(t)$ by letting $\Delta t \to 0$,

$$P_k'(t) + m(t)\cdot P_k(t) = m(t)\cdot P_{k-1}(t) \tag{10}$$

where

$$m(t) = \frac{dg(t)}{dt} \tag{11}$$

A function $m(t)$ is the first derivative or marginal value function of $g(t)$.

To solve the above differential-difference equation (10), we use the method of induction. For k=0, the term on the right hand side becomes zero, so that,

$$P_0'(t) = -m(t)\cdot P_0(t) \tag{12}$$

Integrating both sides of equation (12) with the initial condition, $P_0(0)=1$ yields,

$$P_0(t) = e^{-g(t)} \tag{13}$$

Multiplying both sides of the differential-difference equation (10) by $e^{g(t)}$, we have,

$$e^{g(t)}\{P_k'(t) + m(t)\cdot P_k(t)\} = e^{g(t)}\cdot m(t)\cdot P_{k-1}(t)$$
$$\frac{d\{e^{g(t)}\cdot P_k(t)\}}{dt} = e^{g(t)}\cdot m(t)\cdot P_{k-1}(t) \tag{14}$$

Using the initial condition, $P_0(t)$, for k=1, we have

$$P_1(t) = g(t)\cdot e^{-g(t)} \tag{15}$$

For k=2, we have

$$P_2(t) = \frac{g(t)^2}{2!}\cdot e^{-g(t)} \tag{16}$$

By the method of induction, the following nonstationary Poisson process is derived,

$$P_k(t) = \Pr\{N(t)=k\}$$
$$= \frac{g(t)^k\cdot e^{-g(t)}}{k!} \tag{17}$$

The counting process applied to describe a forest owner's harvesting behavior becomes a Poisson process having parameters, k and $m(t)$ or $\frac{dg(t)}{dt}$. This is the probability that the total number of counts becomes k by time t. If $k=M$, it is the probability that the forest owner's management objective value becomes its threshold by time t.

Waiting Time Process

The primary model given above represents the probability law for the number of counts made by a forest owner in a particular interval of time. Here we further investigate the distribution of the length of time between one count and the next count since a forest owner waits for harvesting forest stands until the management objective value (the number of counts) reaches its threshold.

In order to develop an expression for the distribution of waiting time, let's introduce the waiting time, W_k, until the k-th count occurs, defined by,

$$W_k = \sum_{i=1}^{k} T_i, k \geq 1 \tag{18}$$

where T_i is the interval of time between the (i-1) and the i-th count. If the k-th count will occur prior to or at time τ if and only if the number of counts by time τ is at least k, the distribution function of W_k is derived by,

$$\begin{aligned} F_{W_k}(\tau) &= \Pr\{W_k \leq \tau\} \\ &= \Pr\{N(\tau) \geq k\} \\ &= \sum_{j=k}^{\infty} e^{-g(\tau)} \frac{g(\tau)^j}{j!} \end{aligned} \tag{19}$$

Upon differentiation with respect to time τ, its density function is expressed by,

$$\begin{aligned} \frac{d}{d\tau} F_{W_k}(\tau) &= f_{W_k}(\tau) \\ &= \sum_{j=k}^{\infty} \{-m(\tau) e^{-g(\tau)} \frac{g(\tau)^j}{j!} + m(\tau) e^{-g(\tau)} \frac{g(\tau)^{j-1}}{(j-1)!}\} \\ &= m(\tau) e^{-g(\tau)} \frac{g(\tau)^{k-1}}{(k-1)!} \end{aligned} \tag{20}$$

The Gentan probability, q_i for age class i is the probability that a forest stand is harvested in time interval $[i, i+1)$, therefore it can be derived by integrating the above density function from $\tau=i$ to $\tau=i+1$,

$$\begin{aligned} q_i &= \int_{i}^{i+1} \frac{d}{d\tau} F_{W_k}(\tau) d\tau \\ &= \int_{i}^{i+1} m(\tau) e^{-g(\tau)} \frac{g(\tau)^{k-1}}{(k-1)!} d\tau \end{aligned} \tag{21}$$

If we assume a constant marginal value, λ, for the management objective value function, we have,

$$g(\tau) = \lambda \cdot \tau \tag{22}$$

$$m(\tau) = \frac{dg(\tau)}{d\tau} = \lambda \tag{23}$$

so that the resultant Gentan probability turns out to be an integral of a gamma distribution or Erlang distribution with parameters k and λ,

$$q_i = \int_i^{i+1} \lambda e^{-\lambda\tau} \frac{\lambda \tau^{k-1}}{(k-1)!} d\tau \tag{24}$$

Hypothetical Model Implementation

In this section, an experimental analysis was conducted. In order to simulate a distribution of the Gentan probability, we need to assume the management objective value function with the threshold. The management objective value we used here was based upon the present net value (PNV) of returns from plantation and felling activities. To derive the function, we applied the Richards growth curve (Richards 1958) to project growth of three main species for lumber production in Japan, i.e., Akamatsu (*Pinus densiflora*), Sugi (*Cryptomeria japonica*), and Hinoki (*Chamaecyparis obtusa*). That is, we assume

$$w(t) = M(1 - e^{-at})^b \tag{25}$$

where a, b, and M are coefficients characterizing growth pattern of each, and $w(t)$ is volume per hector at time t. Then we calculate PNV by the following equation.

$$g(t) = PNV(t) = \frac{(P - C_f) \cdot M(1 - e^{-at})^b}{(1+r)^t} - C_r \tag{26}$$

where P, C_f, C_r, and r are a stumpage price per cubic meter, felling costs per cubic meter, regeneration costs per hectare, and discount factor, respectively. Thus we have,

$$\frac{dg(t)}{dt} = m(t)$$

$$= \frac{(P - C_f) \cdot M \cdot a \cdot be^{-at}(1 - e^{-at})^{b-1} - (P - C_f) \cdot M(1 - e^{-at})^b \ln(1+r)}{(1+r)^t} \tag{27}$$

Growth data were obtained from Miyazaki prefecture in Japan. In Table 1, a resultant value for each coefficient of each species is presented, and Figure 2 depicts fitted growth curves.

	M	a	b
Akamatsu	*376.42300	0.04179	1.65315
Pinus densiflora	**23.07260	0.00641	0.18106
	***16.31470	6.51779	9.13044
Sugi	660.95600	0.06745	4.21480
Cryptomeria japonica	2.33626	0.00056	0.04796
	282.91200	119.81000	87.88920
Hinoki	470.77200	0.05163	2.52526
Chamaecyparis obtusa	16.83980	0.00425	0.19870
	27.95590	12.14340	12.70910

*: parameter value, **: standard error, ***: t-value

Table 1. Estimated parameters for the Richards growth function

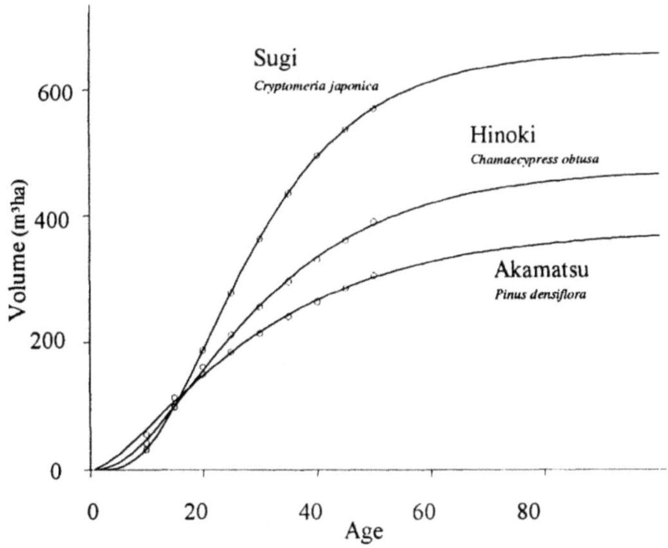

Figure 2. Fitted Richards growth curve

We set the discount factor equal to 3%, and (P, C_f, C_r)=(17,000 Yen/m^3, 8,000Yen/m^3, 0). Also one unit of the management objective value was assumed at 10,000 Yen/ha.

Figure 3 shows the average present net value curves for three species derived from equation (26). Optimal present net value was found at age 28, 35, and 32 for Akamatsu, Sugi, and Hinoki, respectively, which implies that after these ages, the objective value decreases. For the proposed stochastic model, since decrease of the number of counts is not allowed, which is restricted by the assumpion iii), the process terminates before these ages.

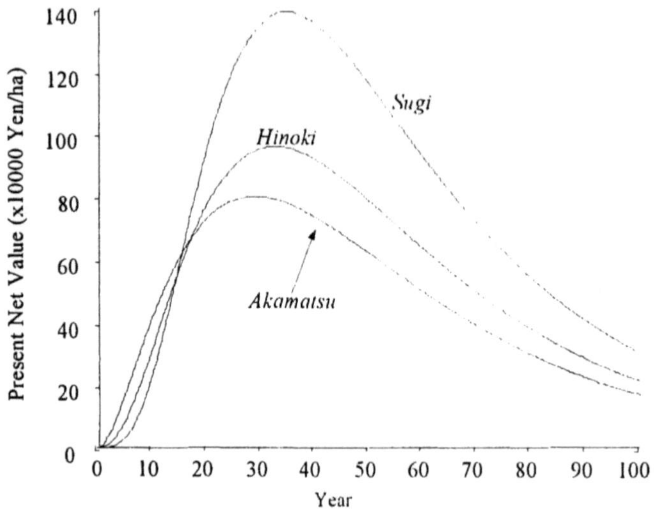

Figure 3. Present net value curve

Using the average present net value function as the management objective function and its derivative, the distribution of the Gentan probability was derived. Figure 4 depicts distributions for each species with different thresholds, M. As can be observed from Figure 4, the more the threshold, the more the distribution moves toward the left-corner, implying waiting longer for harvesting. At the same time, the distribution seems to disappear with the larger threshold. This is true because of the asymptotic nature of the nonstationary management objective value function. The sum of the Gentan probabilities, $\sum_{i=1}^{\infty} q_i$, over time was not always 1 unless a monotonically increasing value function is assumed. The larger the threshold, the less the sum of the probabilities became over time. For instance, when the threshold value was 110 units, the sum of the probabilities was 79%. With the probability of 0.21, a stand could not reach the threshold value. This result is rather appropriate because not all forest stands reach the threshold, allowing some forest owners cease from managing forests. Mathematically speaking, this is because $g(t)$ reaches the point where its derivative becomes zero, i.e. no increase, as t approaches its optimal level. In such a case, no increment of counts is expected, resulting in the sum less than one. Comparing the results in Figure 4 across different species, the sum of the Gentan probabilities for Akamatsu forests tend to decrease earlier than the other two, followed by Hinoki and Sugi. This difference was caused mainly by differences in growth dynamics. Since Sugi grows faster and larger than the others, the threshold can be set at the higher level (see Figure 5).

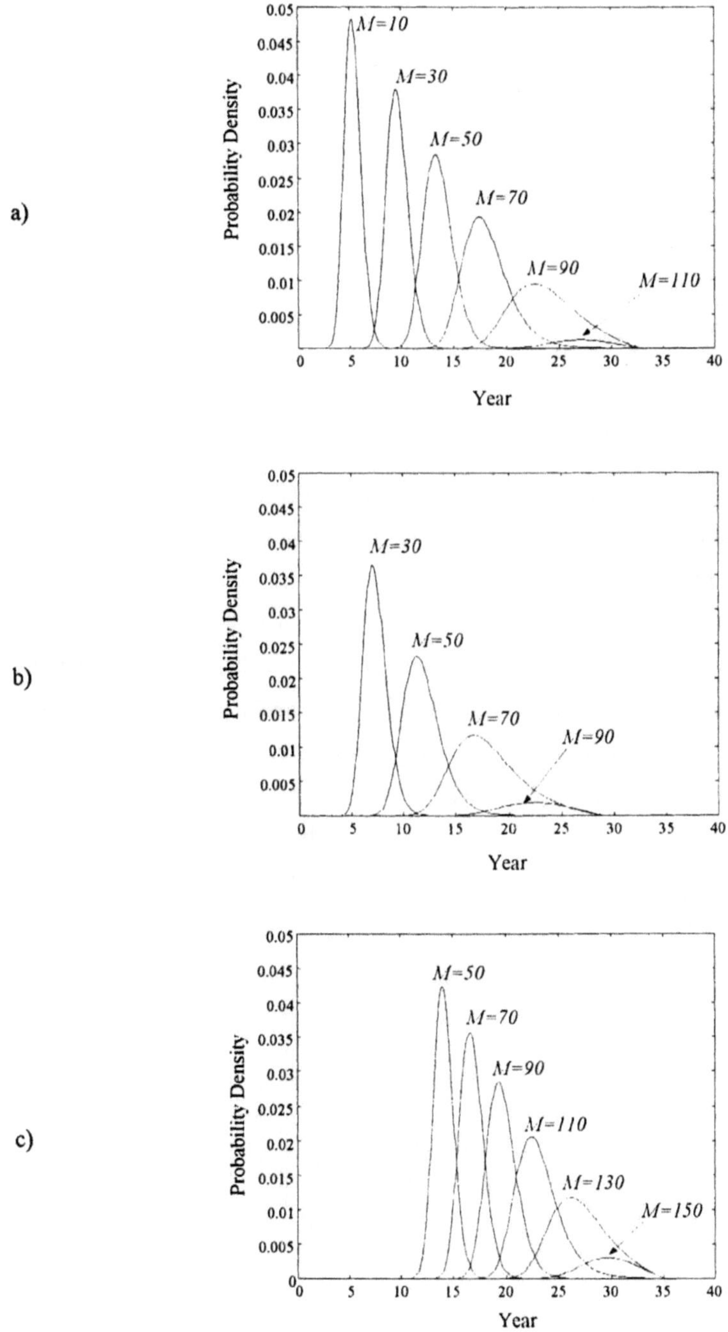

Figure 4. Comparison of Gentan probability density a) Sugi, b) Hinoki, c) Akamatsu

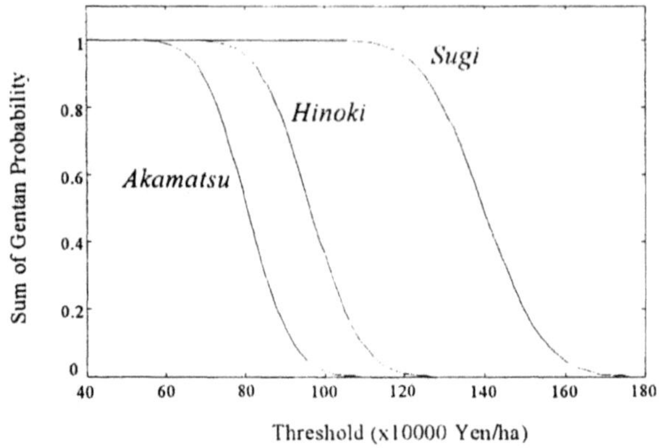

Figure 5. Comparison of the sum of Gentan probabilities

The effect of difference in the threshold value was observed in the felling probabilities. Figure 6 shows the felling probability at different age classes, calculated by equation (5). Three different threshold values, 50, 70 and 90 units, were used for Sugi forest stands. The felling probabilities were estimated over a one year interval. When the threshold was 50 units, harvesting activity was observed from year12 to year 20. On the other hand, harvesting was observed from year 14 to year 24 for the threshold equal to 70 units. The felling probability monotonically increased up to 1. With the threshold equal to 90 units, a large difference exists in the felling probability compared to the others. The felling probability declined after year 25. The first harvesting was observed at age 17, and the last harvesting at age 29. As was observed in Figure 6, the derived felling probabilities yield the longer harvesting time horizon with the larger threshold.

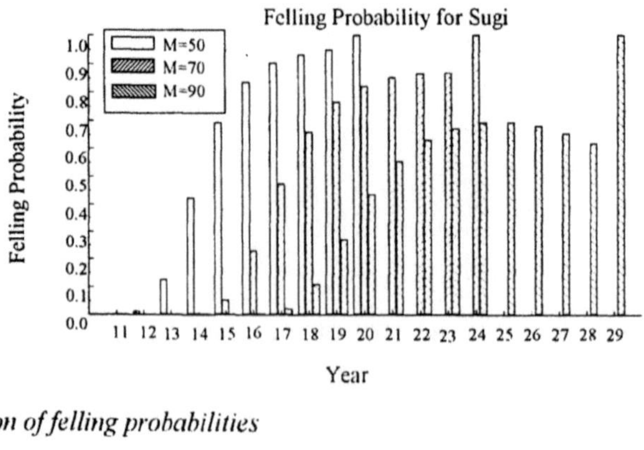

Figure 6. Comparison of felling probabilities

HARVEST DECISIONS UNDER PRICE UNCERTAINTY

In the previous section, we considered a macro-type stochastic model, which deals with timber supply from a large scale of forest resources. We now use a micro-type stochastic model for harvest decision by searching for an optimal rotation age in the forest management framework. Traditionally the total present net value of cash flows from management activities (e.g. plantation, site preparation, thinning, pruning, and final harvest) is usually to be maximized for this purpose. The idea was first introduced by Faustmann (1849), and has been widely accepted to seek financial maturity of a forest stand (see also Samuelson, 1976, Bowes and Krutilla, 1985, Clark, 1990). As can be seen, however, the underlying assumptions for this approach are that all values used in present net value calculation are deterministic or constant over time. These assumptions are often violated due to change in the timber market situation, log prices and various costs, which are uncertain. In the following section we introduce a stochastic model for harvest decision under uncertainty. The proposed model utilized a concept of option valuation.

Stochastic Modeling for Harvest Decision

We assume that log prices follow a stochastic process that evolves through time as the geometric Brownian motion with drift governed by the following stochastic differential equation:

$$
\begin{aligned}
dX_t &= \alpha X_t dt + X_t \sigma dB_t \\
\frac{dX_t}{X_t} &= \alpha dt + \sigma dB_t
\end{aligned}
\qquad (28)
$$

where X_t is a log price at time t, α represents a trend in the relative price, and σ is the instantaneous standard deviation of the relative price change. The term B_t is the standard Brownian motion. The geometric Brownian motion has been commonly used for analyzing stochastic dynamics of stock prices and others because of its simplicity of form and easiness of applying it to the optimization framework. We assume that costs, discount rate as well as forest growth are deterministic.

Given the above continous time stochastic model, the binomial option pricing approximation is used to discretize the process so that the stochastic price is to be incorporated into the discrete framework, i.e. stochastic dynamic programming. Within the binomial option pricing framework, the following parameter transformation yields equation (28) as $\Delta t \to 0$ (Cox et al., 1979);

$$u = e^{\sigma\sqrt{\Delta t}} \qquad (29)$$

$$d = \frac{1}{u} \tag{30}$$

$$P = \frac{e^{\alpha \Delta t} - d}{u - d} \tag{31}$$

where u and d are two possible values for the rate of return on the log price, X_t, with probability P and (1-P). That is, the log price, X_t, increases to $u \bullet X_t$ with probability P, and decreases to $d \bullet X_t$ with probability (1-P) during a time interval (t, $t+\Delta t$). Given the stage interval by unit increment of time, Δt, the future price at the n-th stage with i-times price increment, $S_{n,i}$, becomes;

$$S_{n,i} = u^{i} \cdot d^{n-i-1} \cdot S_{1,0} \tag{32}$$

where $S_{1,0}$ is an initial price equal to X_t at the current stage 1. Equation (32) represents the price dynamics within the discrete framework for equation (28).

The standard discounted value maximization criterion is utilized to seek an optimal decision on management activities within a stochastic dynamic programming framework. Managerial actions or control considered here are to harvest a forest stand now, or not to harvest a forest stand now, i.e. wait now. If a harvesting decision is made, then it becomes necessary to make a decision on whether a forest stand will be replanted or the land converted to alternative uses, i.e. abandoning the management. Figure 7 shows a decision tree based on these actions, given the above binomial process for the log price. Here we use stochastic dynamic programming where time is the stage variable and the stand age and price level are state variables. That is, the proposed model is based on two states and one stage stochastic dynamic programming.

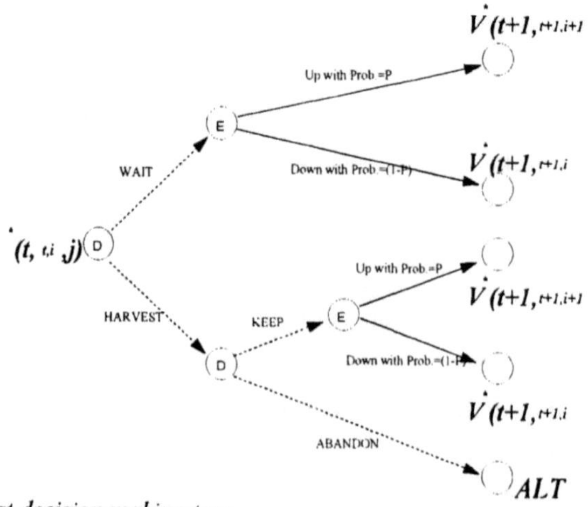

Figure 7. Harvest decision making tree

Let $V^*(t,S_{t,i},j)$ be the objective function to be maximized. This is a discounted value of the total return from activities at stage t, where the forest stand age is j at the i-th state in the binomial process for the log price, $S_{t,i}$. Transition of the state is performed in such a way that if no harvest is implemented at stage t, a forest stand grows by one period. If the log price increases, then a forest stand at the i-th state of the t-th stage moves to the $(i+1)$-th state at the $(t+1)$-th stage, $V^*(t+1,S_{t+1,i+1},j+1)$, while it goes to the i-th state, $V^*(t+1,S_{t+1,i},j+1)$, if the price decreases, i.e., no price increment at this stage. When harvest takes place, given that a forest stand is kept for forest management, a forest stand becomes age 1 at the $(t+1)$-th stage. Depending upon the binomial log price movement, a forest stand goes to $V^*(t+1,S_{t+1,i+1},1)$ if the price goes up, and $V^*(t+1,S_{t+1,i},1)$ with price decrease. If a forest stand management is abandoned for alternative uses, it goes to the abandoned state, ALT. From each state, the decision tree branches off until the end of periods considered. In Figure 7, D represents a node for a decision, that is, the {Wait} or {Harvest} decision, and the {Keep} or {Abandon} decision, while E is for calculation of the expected value of the management using the backward recursion for optimization.

Step 0. Initialization

Set $t=T$, $d_{T,i,j}=\{H\text{-}A\}$ for $i=0,,,T\text{-}1$ and $j=1,,,J+T\text{-}1$.

Calculate $V^*(T,S_{T,i},j)=S_{T,i}\bullet Q_j+ALT$ for $i=0,1,,,T\text{-}1$, $j=1,2,,J+T\text{-}1$

given that $S_{T,i} = u^i \cdot d^{T-i-1} \cdot S_{1,0}$

Step 1. Iteration

Set $t=t\text{-}1$. Find an optimal decision, $d^*_{t,i,j}(\in D_t)$for $i=0,,,t\text{-}1$ and $j=1,,,J+t\text{-}1$ such that

$$V^*(t,S_{t,i},j) = \max[V(t,S_{t,i},j,\{W\}), V(t,S_{t,i},j,\{H-K\}), V(t,S_{t,i},j,\{H-A\})]$$

where

$$V(t,S_{t,i},j,\{W\}) = \frac{P \cdot V^*(t+1,S_{t+1,i+1},j+1)+(1-P)\cdot V^*(t+1,S_{t+1,i},j+1)-C_a}{1+r}$$

$$V(t,S_{t,i},j,\{H-K\}) = S_{t,i}\cdot Q_j + \frac{P\cdot V^*(t+1,S_{t+1,i+1},1)+(1-P)\cdot V^*(t+1,S_{t+1,i},1)-C_a}{1+r} - C_r$$

$$V(t,S_{t,i},j,\{H-A\}) = S_{t,i}\cdot Q_j + ALT$$

given that $S_{t,i} = u^i \cdot d^{t-i-1} \cdot S_{1,0}$

Step 2. Iterative Judgement

If $t=1$, then go to Step 3, otherwise go to Step 1.

Step 3. Final Judgement

Find an optimal rotation age, t_{opt} such that

$$t_{opt} = \{j | \inf j \textit{ subject to } d^*_{1,0,j} = [\{H-K\} \textit{ or } \{H-A\}]\}$$

Then terminate the procedure.

Figure 8. Recursive procedure for the proposed stochastic dynamic programming model

At the stage t, the expected value along with the {Wait} decision made for the stand management with age j and i-times price increment, $E_{t,i,j}[W]$, is expressed by

$$E_{t,i,j}[W]=\frac{P\cdot V^{*}(t+1,S_{t+1,i+1},j+1)+(1-P)\cdot V^{*}(t+1,S_{t+1,i},j+1)-C_{a}}{1+r} \tag{33}$$

where r is a constant one period discount rate and C_a is an annual cost for management. On the other hand, if the {Harvest} decision is made, calculation of the expected value becomes dependent upon another decision, {Keep} or {Abandon}. This decision is also based on the expected value from the corresponding outcomes. That is, the expected value under the {Harvest} decision, $E_{t,i,j}[H]$, at the t-th stage with i-times increment for the j age stand becomes

$$E_{t,i,j}[H]=S_{t,i}\cdot Q_{j}+\max\{ALT,\frac{P\cdot V^{*}(t+1,S_{t+1,i+1},1)+(1-P)\cdot V^{*}(t+1,S_{t+1,i},1)-C_{a}}{1+r}-C_{r}\} \tag{34}$$

where $S_{t,i}$ is the log price at the i-th state (i-times increment) of the t-th stage, Q_j is a forest stand volume at age j, and C_r is a regeneration cost, which is charged at stage t just after harvesting. The first term on the right hand side, $S_{t,i} \bullet Q_j$ is the return from harvesting the stand. The second term with „max" is the value of the management after harvesting. If an alternative use has a higher value than the discounted expected value of managing the stand, then the land is converted into the alternative use, and forest management is abandoned. Otherwise, the forest stand is kept for management. As can be seen in equations (33) and (34), the expected value of the {Wait} and {Keep} decision is discounted by one period. An optimal decision at each state of each stage is searched by

$$V^{*}(t,S_{t,i},j)=\max\{E_{t,i,j}[W],E_{t,i,j}[H]\} \tag{35}$$

The binomial decision tree branches from t=1 to T with i=0 to t-1 and j=1 to J+t-1, where T is the last stage for optimization and J is the age of the oldest forest stand at the current stage 1. The number of states and the oldest forest stand age increases by one with increase in the stage.

To seek an optimal decision at the current stage, $V^{*}(1,S_{1,0},j)$ for j=1 to J, the two-state and one-stage stochastic dynamic programming technique is utilized with the backward recursive procedure. Recursive searching procedure begins from the last stage up to the current stage. The {Abandon} decision is to be selected at first for each forest stand at each state of the last stage. Given a forest age range from 1 to J at the first or current stage, the (J+T-1) different age classes of forest stands can be observed at the last stage, T. For these forests at the last stage, the decision of {Harvest}-{Abandon} is to be made first under all states of price level, $\{S_{T,0}, S_{T,1}, S_{T,2},.., S_{T,T-1}\}$.

Let D_t be the decision set at stage t consisting {W}, {H-K} and {H-A} that represent the {Wait} decision, the {Harvest}-{Keep} decision, and the {Harvest}-{Abandon} decision, respectively. Also let $d^*_{t,i,j}$ be the optimal decision at the (i,j) state of stage t. At the last stage T, thus we have $d^*_{T,i,j}$={H-A} for i=0,,,T-1 and j=1,,,J+T-1. The corresponding objective value from activities is

$$V^*(T, S_{T,j}, j) = S_{T,i} \cdot Q_j + ALT \ for \ i = 0,1,,,T-1, j = 1,2,,J+T-1 \quad (36)$$

where ALT denotes value of alternative uses. The backward recursive procedure in the proposed two-state and one-stage stochastic dynamic programming starts as described in Figure 8. As can be seen, once $V^*(1,S_{1,0},j)$ for all j is searched, the youngest stand age with the {Harvest}-{Keep} decision at the current stage is regarded as an optimal rotation age, given the current log price, $S_{1,0}$, along with the corresponding expected present net value.

Coefficient Estimation for Stochastic Differential Equation

In order to search for an optimal rotation age with the proposed log price stochastic model under the stochastic dynamic programming, coefficients of the proposed stochastic differential equation (28) have to be estimated first. We utilized the maximum likelihood method for coefficient estimation. It was carried out by discretizing the process in the following way. Given a log price, X_t, we introduce a new variable y_t , where $y_t = f(X_t)$ = $\log(X_t)$. Applying Ito's formula to y_t, we have

$$dy_t = f'(X_t)dX_t + \frac{1}{2}f''(X_t)(dX_t \cdot dX_t) \quad (37)$$

Since the first and second derivatives of y_t are

$$\frac{dy_t}{dX_t} = f'(X_t) = \frac{1}{X_t} and \frac{d^2y_t}{dX_t^2} = f''(X_t) = -\frac{1}{X_t^2} \quad (38)$$

equation (37) becomes

$$dy_t = \frac{1}{X_t}dX_t + \frac{1}{2}(-\frac{1}{X_t^2})(dX_t \cdot dX_t) \quad (39)$$

Replacing dX_t by equation (28), equation (39) becomes

$$\begin{aligned} dy_t &= \frac{1}{X_t}(\alpha X_t dt + \sigma X_t dB_t) + \frac{1}{2}(-\frac{1}{X_t^2})(\alpha X_t dt + \sigma X_t dB_t)(\alpha X_t dt + \sigma X_t dB_t) \\ &= \alpha dt + \sigma dB_t - \frac{1}{2}\{\alpha^2(dt \cdot dt) + \sigma^2(dB_t \cdot dB_t) + 2\alpha\sigma dt \cdot dB_t\} \end{aligned} \quad (40)$$

Since $(dt \bullet dt) = dt \bullet dB_t = dB_t \bullet dt = 0$ and $(dB_t \bullet dB_t) = dt$ (see Oksendal, 1995), equation (40) further becomes

$$dy_t = (\alpha - \frac{1}{2}\sigma^2)dt + \sigma dB_t \quad (41)$$

Notice that there is no variable, y_t, on the right hand side, implying that it becomes the simple

Brownian motion with drift.

For estimation with discrete price data, equation (41) is discretized by replacing dy_t by $(y_{t_{n+1}} - y_{t_n})$

$$\begin{aligned} y_{t_{n+1}} - y_{t_n} &= (\alpha - \frac{1}{2}\sigma^2)(t_{n+1} - t_n) + \sigma(B_{t_{n+1}} - B_{t_n}) \\ \text{or} \quad y_{t_{n+1}} &= y_{t_n} + (\alpha - \frac{1}{2}\sigma^2)(t_{n+1} - t_n) + \sigma(B_{t_{n+1}} - B_{t_n}) \end{aligned} \tag{42}$$

Since B_{t_n} is the standard Brownian motion, then $(B_{t_{n+1}} - B_{t_n})$ follows the normal distribution law with a zero mean and a variance equal to $(t_{n+1}-t_n)$. Setting $\varepsilon_n = (B_{t_{n+1}} - B_{t_n})$, equation (42) becomes

$$y_{t_{n+1}} = y_{t_n} + (\alpha - \frac{1}{2}\sigma^2)(t_{n+1} - t_n) + \sigma\varepsilon_n \tag{43}$$

resulting in

$$\varepsilon_n = \frac{y_{t_{n+1}} - y_{t_n} - (\alpha - \frac{1}{2}\sigma^2)(t_{n+1} - t_n)}{\sigma} \tag{44}$$

where

$$\varepsilon_n \sim N(0, t_{n+1} - t_n) \tag{45}$$

Given a set of data, $(X_{t_0}, X_{t_1}, X_{t_2},,, X_{t_N})$, the maximum likelihood estimates of the parameters (α, σ) are estimated. The likelihood function for the sample set $(X_{t_0}, X_{t_1}, X_{t_2},,, X_{t_N})$ is expressed by

$$p(X_{t_0}, X_{t_1}, X_{t_2},,, X_{t_N}) = p(y_{t_0}, y_{t_1}, y_{t_2},,, y_{t_N}) \left| \frac{\partial(y_{t_0}, y_{t_1}, y_{t_2},,, y_{t_N})}{\partial(X_{t_0}, X_{t_1}, X_{t_2},,, X_{t_N})} \right| \tag{46}$$

using the transformed sample set $(y_{t_0}, y_{t_1}, y_{t_2},,, y_{t_N})$, where the last element of equation (46), $|\bullet|$, denotes the absolute value of the determinant of the Jacobian matrix. Taking the logarithm for both sides of equation (46), the log-likelihood function based on the observation $(X_{t_0}, X_{t_1}, X_{t_2},,, X_{t_N})$ becomes

$$\begin{aligned} \log p(X_{t_0}, X_{t_1}, X_{t_2},,, X_{t_N}) &= \log p(y_{t_0}, y_{t_1}, y_{t_2},,, y_{t_N}) + \sum_{n=0}^{N} \log(\frac{1}{X_{t_n}}) \\ &= \log p(y_{t_0}, y_{t_1}, y_{t_2},,, y_{t_N}) - \sum_{n=0}^{N} \log(X_{t_n}) \end{aligned} \tag{47}$$

That is, given the observation, $(X_{t_0}, X_{t_1}, X_{t_2},,, X_{t_N})$, the values of the parameter set is found by maximizing equation (47), where the likelihood function for the transformed sample set

$(y_{t_0}, y_{t_1}, y_{t_2}, ,, y_{t_N})$ on the right hand side is expressed by

$$p(y_{t_0}, y_{t_1}, y_{t_2}, ,, y_{t_N}) = p(y_{t_0}) \cdot \prod_{n=0}^{N-1} p(y_{t_{n+1}} \mid y_{t_n}) \tag{48}$$

Since $\{\varepsilon_n\}$ follows $N(0, t_{n+1}-t_n)$, thus from equation (44) we have

$$p(y_{t_{n+1}} \mid y_{t_n}) = \frac{1}{\sqrt{2\pi\sigma^2 (t_{n+1} - t_n)}} \exp\left[-\frac{1}{2} \cdot \frac{\{y_{t_{n+1}} - y_{t_n} - (\alpha - \frac{\sigma^2}{2})(t_{n+1} - t_n)\}^2}{\sigma^2 (t_{n+1} - t_n)} \right] \tag{49}$$

The log-likelihood function for $(y_{t_0}, y_{t_1}, y_{t_2}, ,, y_{t_N})$ becomes

$$\begin{aligned} &\log p(y_{t_0}, y_{t_1}, y_{t_2}, ,, y_{t_N}) \\ &= \sum_{n=0}^{N-1} \log p(y_{t_{n+1}} \mid y_{t_n}) + \log p(y_{t_0}) \\ &= -\frac{1}{2} \sum_{n=0}^{N-1} \left[\log\{2\pi\sigma^2 (t_{n+1} - t_n)\} + \frac{\{y_{t_{n+1}} - y_{t_n} - (\alpha - \frac{\sigma^2}{2})(t_{n+1} - t_n)\}^2}{\sigma^2 (t_{n+1} - t_n)} \right] + \log p(y_{t_0}) \end{aligned} \tag{50}$$

As a result, using both sample sets $\{ y_{t_n} \}$ and $\{ X_{t_n} \}$ and equation (47), the maximum likelihood estimates of the parameters (α, σ) are calculated by maximizing

$$\begin{aligned} \log p(X_{t_0}, X_{t_1}, X_{t_2}, ,, X_{t_N}) = &-\frac{1}{2} \sum_{n=0}^{N-1} \left[\log\{2\pi\sigma^2 (t_{n+1} - t_n)\} + \frac{\{y_{t_{n+1}} - y_{t_n} - (\alpha - \frac{\sigma^2}{2})(t_{n+1} - t_n)\}^2}{\sigma^2 (t_{n+1} - t_n)} \right] \\ &+ \log p(y_{t_0}) - \sum_{n=0}^{N} \log(X_{t_n}) \end{aligned} \tag{51}$$

Given the observation on the log price, the parameters (α, σ) of the stochastic differential equation (28) are estimated by the above maximum likelihood method. Once (α, σ) are estimated with the corresponding Δt, the parameters (u, d, P) for the binomial option pricing approximation are obtained, and used in the proposed two-state and one-stage stochastic dynamic programming to search for an optimal rotation age.

Optimal Rotation under Price Uncertainty

For the analysis conducted here, we used log prices of Akamatsu (*Pinus densiflora*), Sugi (*Cryptomeria japonica*) and Hinoki (*Chamaecyparis obtusa*) at the Japanese timber market. The data are monthly data. The data source was "Report of Supply and Demand of Timber" from the Japanese Forestry Agency.

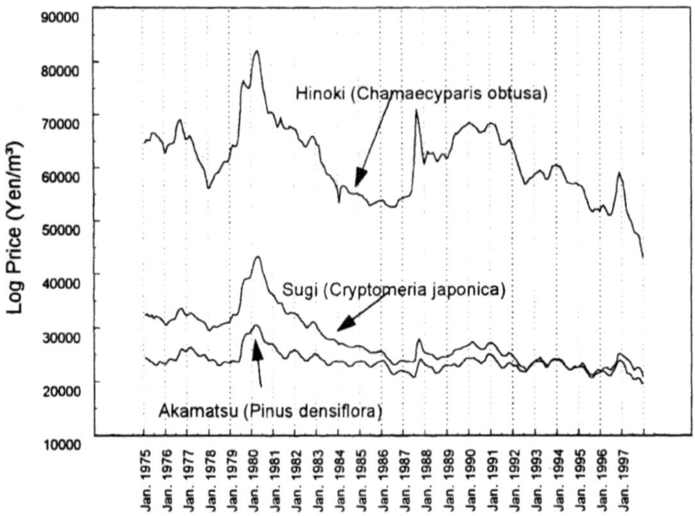

Figure 9. Log price dynamics for three species

Figure 9 shows price trends of Akamatsu, Sugi and Hinoki from 1975 to 1997. The large price change observed in 1980 was most likely due to the second Oil shock. For growth projection of these species, we used the result in Table 1. The annual management cost was set to 30,000 Yen/ha, while the regeneration cost was assumed 300,000 Yen/ha. A one percent interest rate was used. The value for the alternative use was assumed to be zero, implying that if the value of management becomes nonpositive, then the {Abandon} decision is selected. With the monthly price data, discretizing the continuous time stochastic model was conducted with a time interval, $(t_{n+1}-t_n)=\Delta t=1/12$. With $\Delta t=1/12$, the maximum likelihood method was utilized to estimate coefficients. Results of estimation by the maximum likelihood method are presented in Table 2. For three species, t-statistics of a coefficient α became low, resulting in rejection of the $\{\alpha \neq 0\}$ hypothesis with a 5% significance level. Thus, we used zero and non-zero value for α for re-estimation of σ.

A two-state and one-stage stochastic dynamic programming model for the proposed binomial decision tree was built in order to search for an optimal rotation age under stochastic prices. Since searching was conducted on a yearly basis, we set the time interval to 1 year, i.e., $\Delta t=1$ for discretization for the binomial approximation. For comparison purposes, solution from the Faustmann present net value approach under the deterministic price was also derived by applying $u=d=1$ with $P=1.00$, i.e. a constant price over time in the following manner,

$$V^F(t,j) = \max\{E^F_{t,j}[W], E^F_{t,j}[H]\} \tag{52}$$

where

$$E^F_{t,j}[W] = \frac{V^F(t+1,j+1) - C_a}{1+r} \tag{53}$$

and

$$E^F_{t,j}[H] = S_{1,0} \cdot Q_j + \frac{V^F(t+1,1) - C_a}{1+r} - C_r \tag{54}$$

$V^F(t,j)$ is a present net value of the total return from activities for the j-th age forest stand at the t-th stage. $E^F_{t,j}[W]$ and $E^F_{t,j}[H]$ are the expected values with the {Wait} and {Harvest} decision at stage t for a stand with age j, respectively.

Figure 10 shows the relationship between the current log price and the derived optimal rotation age. In this case, zero value was applied to α of equation (28). The solid line in Figure 10 was for the value derived under stochastic prices, while the dot line was from the Faustmann deterministic approach. The current log price was set from 17,000 Yen/m^3 to 47,000 Yen/m^3 for three species with 8,000 Yen/m^3 as harvesting costs, and 300,000 Yen/ha as regeneration costs. As was observed in Figure 10, when the price for Sugi was within the range from 17,000 to 18,600 Yen/m^3 in all age classes, the {Harvest}-{Abandon} decision was selected, implying that maintaining forest management resulted in a negative present net value. For Hinoki and Akamatsu, that range was 17,000 to 19,500 Yen/m^3 and 17,000 to 19,900 Yen/m^3, respectively. The {Harvest}-{Abandon} decision was selected up to the point where the log price became 19,400 Yen/m^3 for Sugi, 20,800 Yen/m^3 for Hinoki, and 21,700 Yen/m^3 with increase in the optimal rotation age or the earliest time to abandon. The resultant optimal age to abandon management increased to 66, 82 , and 66 years for Sugi, Hinoki, and Akamatsu, respectively. After that peak, the {Harvest}-{Keep} decision was selected, and the resultant optimal rotation age decreased to 33, 29 and 20 years for Sugi, Hinoki and Akamatsu along with price increase up to 47,000 Yen/m^3. These differences were reflected by difference in growth projection of each species. For fast growing species like Sugi, the management can be kept under low prices, compared to the others. Slow growing Akamatsu needs high log prices to maintain the management.

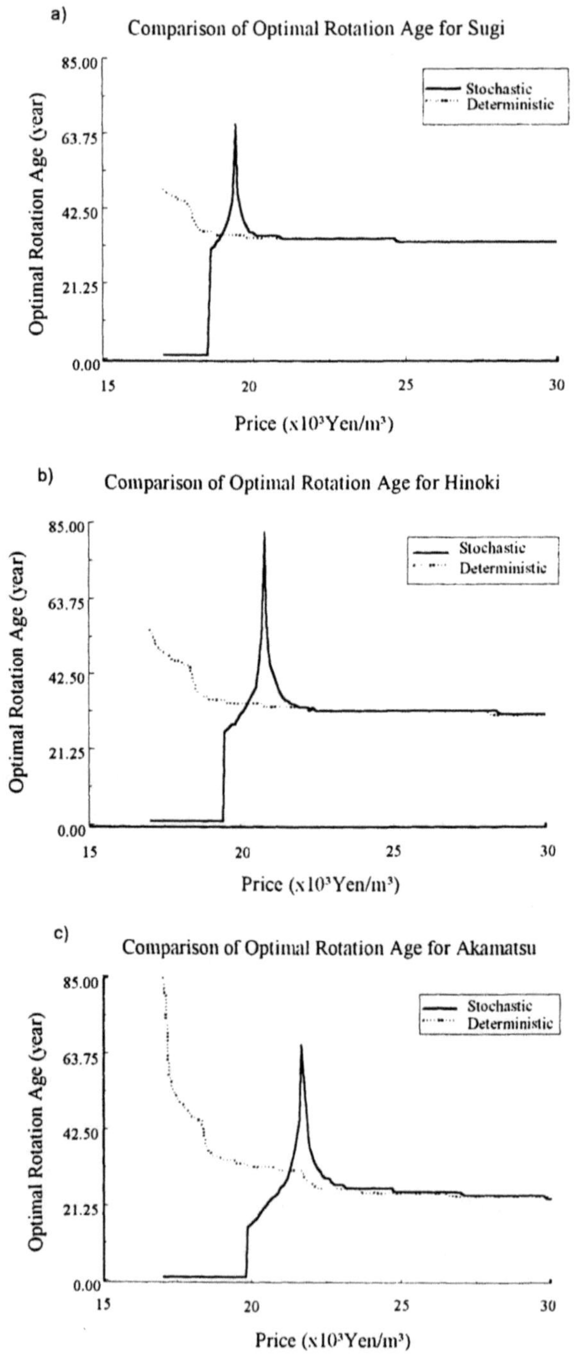

Figure 10. Comparison of stochastic and deterministic models a) Sugi, b) Hinoki, c) Akamatsu

The results from the Faustmann approach for a Sugi stand, on the other hand, showed that the optimal rotation age decreased from 48 years to 33 years with increase in the current log price, while for Hinoki it decreased from 55 to 31 years and for Akamatsu from 85 to 20 years. The difference in the two approaches, stochastic and Faustmann, can be observed during the period when {Abandon} and {Keep} decisions are exchanged. When non-zero value was applied to α, an optimal rotation age and timing to switch from the {Abandon} decision to the {Keep} decision changed as shown in Figure 11. For three species, α was negative, implying more probability for the price to decrease. This pushed the price higher to have a nonnegative objective value. For Sugi, such a price changed from 18,600 to 20,700 Yen/m^3 for the first nonnegative objective value, and from 19,500 to 21,400 Yen/m^3 for Hinoki, and from 19,900 to 20,400 Yen/m^3 for Akamatsu. Because of the different negative value of α, -0.02055 for Sugi, -0.01614 for Hinoki, and –0.00463 for Akamatsu, the largest change in the prices was observed for Sugi, followed by Hinoki and Akamatsu.

The {Harvest}-{Abandon} decision was kept until the price became 23,200 Yen/m^3 with optimal rotation 33 years for Sugi, decreasing to 28 years as the price further increased. For Hinoki, when the optimal rotation reached its peak at age 33, the price was 24,600 Yen/m^3, then the optimal rotation decreased to 24 years. Because of a small value for α, Akamatsu showed less changes than the other two. Forty-three years was the longest optimal rotation age at a price equal to 22,900 Yen/m^3. After its peak, the optimal rotation approached 19 years.

Differences in the solutions for both zero and non-zero values for α posed a question if the proposed continuous time stochastic model can describe price uncertainty effectively. As stated earlier, simplicity is a big advantage to utilize the geometric Brownian motion. Estimation difficulty has prevented us from applying more complicated models. Since a slight change in estimated coefficients brings a large change in outcomes, more candidates for the model are necessary for a final decision on the model selection. In the next section, the local linearization technique is presented to estimate coefficients of nonlinear stochastic differential equations.

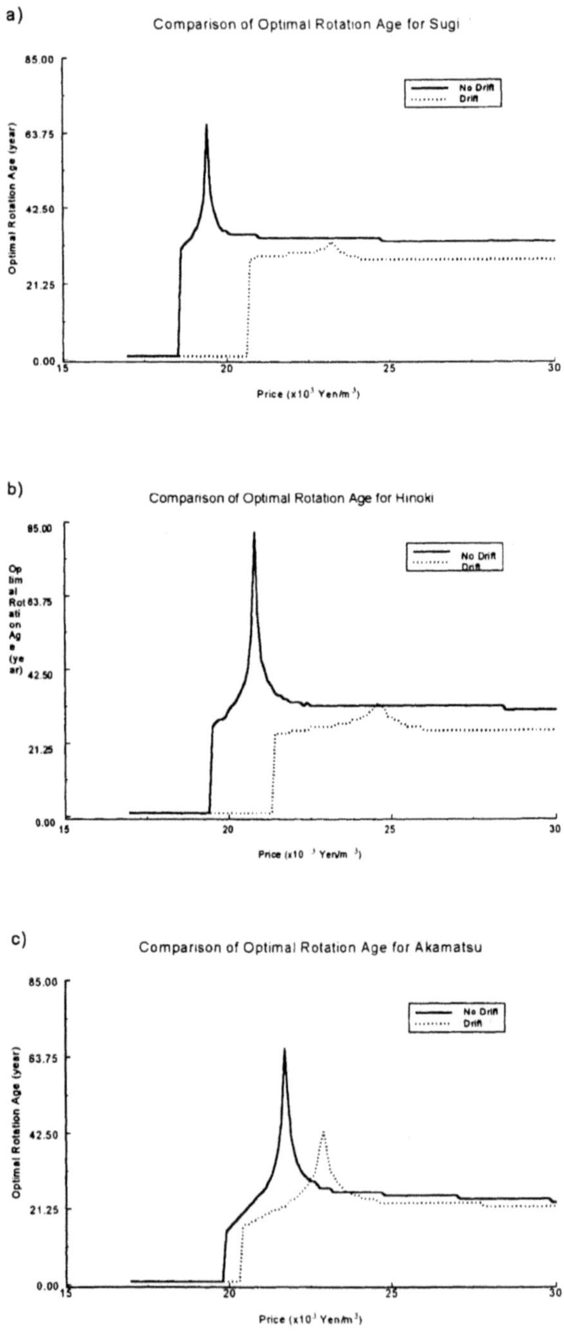

Figure 11. Comparison of two derived stochastic models. a) Sugi, b) Hinoki, c) Akamatsu

STOCHASTIC MODEL SPECIFICATION AND ESTIMATION

In this section, focusing on log price stochastic behavior, we propose thirteen alternatives for a continuous time stochastic model based on the state dependent volatility process (SDVP), then based on AIC (Akaike, 1977) we investigate if there exists a difference in the performance of each model. We apply the local linearization method to estimate parameters of models.

The stochastic models we considered here are based on the following SDVP,

$$dx_t = (\alpha + \beta x_t)dt + x^{\gamma}\sigma dB_t \tag{55}$$

where x_t is a log price, B_t is a standard Brownian motion, and α, β, and γ are coefficients. This is a stochastic differential equation with a drift term, which is a linear function of x_t. In order to propose thirteen different SDVPs, we used the tree diagram described in Figure 12.

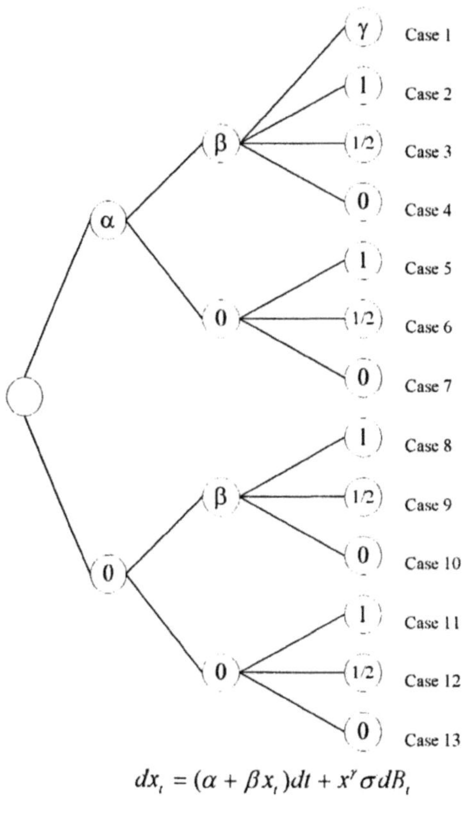

Figure 12. Tree diagram for model generation

Changing restriction on coefficients based on the tree diagram in Figure 12, the thirteen different SDVPs were constructed as in Table 3 in the following way. At the first branch on the tree diagram, a constant drift term, α, is determined to be either zero or unrestricted, while at the second branch a state dependent drift term, β, is set to either zero or unrestricted. At the last branch, a power coefficient of x_t, is branched to be 1, 1/2 and zero from each node. The model of Case 1 is added as a general form of SDVP, unrestricting a power coefficient, γ.

Case #	Stochastic Differential Equation
Case 1	$dx_t = (\alpha + \beta x_t)dt + x^{\gamma}\sigma dB_t$
Case 2	$dx_t = (\alpha + \beta x_t)dt + x\sigma dB_t$
Case 3	$dx_t = (\alpha + \beta x_t)dt + x^{1/2}\sigma dB_t$
Case 4	$dx_t = (\alpha + \beta x_t)dt + \sigma dB_t$
Case 5	$dx_t = \alpha dt + x\sigma dB_t$
Case 6	$dx_t = \alpha dt + x^{1/2}\sigma dB_t$
Case 7	$dx_t = \alpha dt + \sigma dB_t$
Case 8	$dx_t = \beta x_t dt + x\sigma dB_t$
Case 9	$dx_t = \beta x_t dt + x^{1/2}\sigma dB_t$
Case 10	$dx_t = \beta x_t dt + \sigma dB_t$
Case 11	$dx_t = x\sigma dB_t$
Case 12	$dx_t = x^{1/2}\sigma dB_t$
Case 13	$dx_t = \sigma dB_t$

Table 3. Candidates for State Dependent Volatility Process

Local Linearization Method

Parameter estimation for each model was carried out by using the maximum likelihood method with the local linearization method. The idea of the local linearization method was first introduced by Ozaki (1985). It has been paid a great deal of attention for parameter estimation of nonlinear stochastic differential equations. Recent extension of the method was implemented by Shoji and Ozaki (1997). The basic idea of the method is that an original stochastic differential equation is locally approximated by a linear stochastic differential equation. In what follows, we review the method.

Let's assume the following general form of the state dependent stochastic process,

$$dx_t = f(x_t)dt + g(x_t)dB_t \tag{56}$$

where $f(x_t)$ and $g(x_t)$ are twice continuously differentiable with respect to x_t and B_t is a standard Brownian motion. Using Ito's formula with the following transformation,

$$y_t = \phi(x_t) \tag{57}$$

$$\phi'(x_t) \cdot g(x_t) = \sigma \tag{58}$$

we have,

$$dy_t = h(y_t)dt + \sigma dB_t \tag{59}$$

where

$$h(y_t) = \{\frac{1}{2} g(x_t)^2 \cdot \phi''(x_t) \cdot \sigma^2 + f(x_t) \cdot \phi'(x_t)\} \tag{60}$$

The local linearization method is based on the assumption that $h(y_t)$ is locally a linear function of y_t. Under Ito's formula $h(y_t)$ becomes

$$dh(y_t) = h'(y_t)dy_t + \frac{\sigma^2}{2} h''(y_t)dt \tag{61}$$

For small time interval $[t, t+\Delta t)$, it can be assumed that $h'(y_t)$ and $h''(y_t)$ are constant, and thus

$$h(y_s) = h(y_t) + h'(y_t)(y_s - y_t) + \frac{\sigma^2}{2} h''(y_t)(s - t) \tag{62}$$

for $s \in [t, t+\Delta t)$. Applying this approximation into equation (59), we have

$$dy_t = (L_t y_s + M_t s + N_t)ds + \sigma dB_s \tag{63}$$

where

$$L_t = h'(y_t) \tag{64}$$

$$M_t = \frac{\sigma^2}{2} h''(y_t) \tag{65}$$

$$N_t = h(y_t) - h'(y_t)y_t - \frac{\sigma^2}{2} h''(y_t)t \tag{66}$$

Solving equation (63) yields, in the discrete sense

$$\begin{aligned} y_{t+\Delta t} = \ & y_t + \frac{h(x_t)}{L_t}(\exp(L_t \Delta t) - 1) + \frac{M_t}{L_t^2}(\exp(L_t \Delta t) - 1 - L_t \Delta t) \\ & + \sigma \int_t^{t+\Delta t} \exp(L_t(t + \Delta t - s))dB_s \end{aligned} \tag{67}$$

It is noticeable that

$$\sigma \int_t^{t+\Delta t} \exp(L_t(t + \Delta t - s))dB_s \tag{68}$$

on the right hand side follows normal distribution with mean 0 and variance, V_t,

$$V_t = \frac{\sigma^2 \cdot (\exp(2L_t \Delta t) - 1)}{2L_t} \tag{69}$$

Thus, parameters of equation (56) can be obtained by maximizing the log-likelihood function of $\{y_t\}$, which is

$$\log(p(y_{t_0}, y_{t_1}, \cdots, y_{t_N})) = -\frac{1}{2}\sum_{i=1}^{N}\{\frac{(y_{t_i} - E_{t_{i-1}})^2}{V_{t_{i-1}}} + \log(2\pi V_{t_{i-1}})\} + \log(p(y_{t_0})) \quad (70)$$

where

$$E_t = y_t + \frac{h(y_t)}{L_t}(\exp(L_t\Delta t) - 1) + \frac{M_t}{L_t^2}(\exp(L_t\Delta t) - 1 - L_t\Delta t) \quad (71)$$

For the original observations $\{ x_t \}$,

$$p(x_{t_0}, x_{t_1}, \cdots, x_{t_N}) = p(y_{t_0}, y_{t_1}, \cdots, y_{t_N})\left|\frac{\partial(y_{t_0}, y_{t_1}, \cdots, y_{t_N})}{\partial(x_{t_0}, x_{t_1}, \cdots, x_{t_N})}\right| \quad (72)$$

so that the log-likelihood function becomes

$$\log(p(x_{t_0}, x_{t_1}, \cdots, x_{t_N})) = \log(p(y_{t_0}, y_{t_1}, \cdots, y_{t_N})) + \sum_{i=0}^{N}\log\{\frac{d\phi(x_{t_i})}{dx_{t_i}}\} \quad (73)$$

where the last element of equation (72), $|\bullet|$, denotes the absolute value of the determinant of the Jacobian matrix.

Comparison of Model Performance

In our experiments, log prices of the following thirteen species at the Japanese timber market were used on a monthly basis from January 1975 to June 1994. They are Akamatsu (*Pinus densiflora*), Sugi (*Cryptomeria japonica*), Hinoki (*Chamaecyparis obtusa*), Karamatsu (*Larix leptolepis*), Ezomatsu (*Pinaceae jezoensis*), Nara (*Quercuscrispula*), Buna (*Fagus crenata*), Rawan (*Dipterocar paceae*), Douglas-fir (*Pseudotsuga menziesii*}), Hemlock (*Tsuga heterophylla*), Russian Spruce, Russian Larch (*Larix sibirica*), and Radiata Pine (*Pinus radiata*). The first seven are domestic and the rest are imported species of forest products in Japan. The data source was "Report of Supply and Demand of Timber" from the Japanese Forestry Agency. With the base month equal to January 1975, Figure 13-a shows the log price index for the domestic logs, and Figure 13-b is for the six imported logs. A large price change was observed in 1981 most likely due to the second Oil shock.

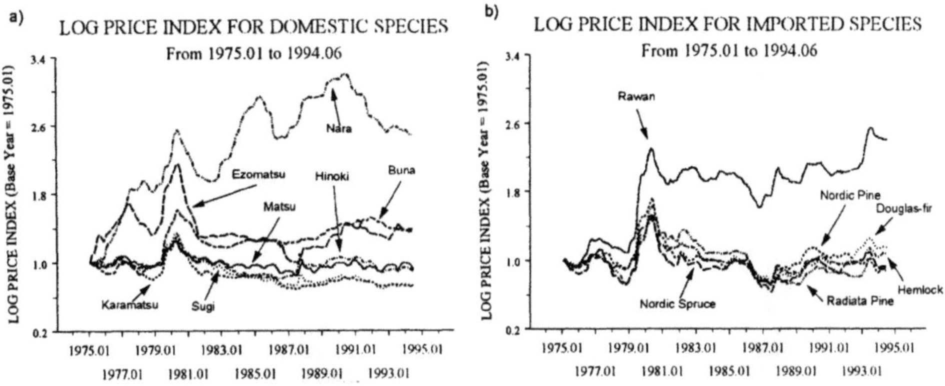

Figure 13. Log price indices a) Domestic species, b) Imported species

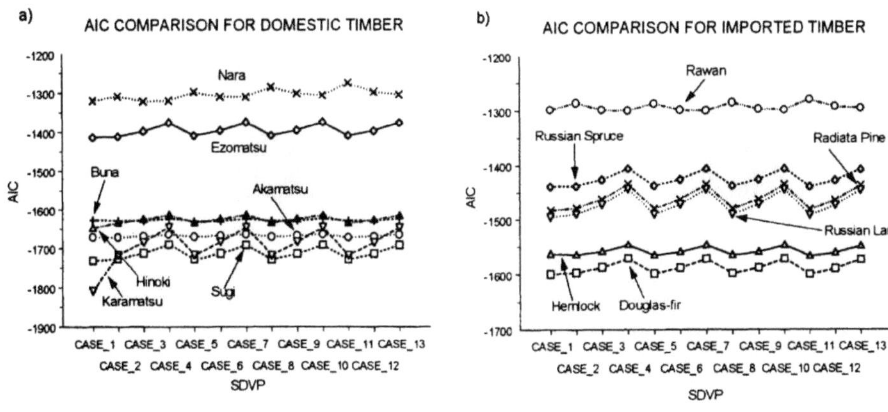

Figure 14. Comparison of AIC over different models a) Domestic species, b) Imported species

Parameters of SDVP's in Table 3 were derived by maximizing (73). Figure 14 shows the resultant AIC for each stochastic model in Table 3. As for the domestic logs, there is a tendency for species on AIC results (Figure 14-a). Sugi, Hinoki, Karamatsu, and Ezomatsu have the same pattern on AIC over the different models. They have high AIC for Case 4, 7, 10 and 13, where the stochastic model has a constant diffusion coefficient, and low AIC for Case 1, 5, 8 and 11 with a state dependent diffusion coefficient, $x_t\sigma$. Nara has almost an opposite tendency, i.e. high AIC for Case 2, 5, 8 and 11, and low AIC for Case 4, 7, 10 and 13. The rest, Matsu and Buna, showed almost no difference in AIC. As for the imported logs in Figure 14-b, all but Rawan have the same tendency on AIC as Sugi, Hinoki, Karamatsu and Ezomatsu. The stochastic model with a constant diffusion coefficient tends to have high AIC and the model with a state dependent diffusion coefficient shows low AIC. Rawan showed the same pattern on AIC as Nara. Overall, a general form of SDVP, Case 1 yielded the best AIC perfor3mance, followed by Case 11.

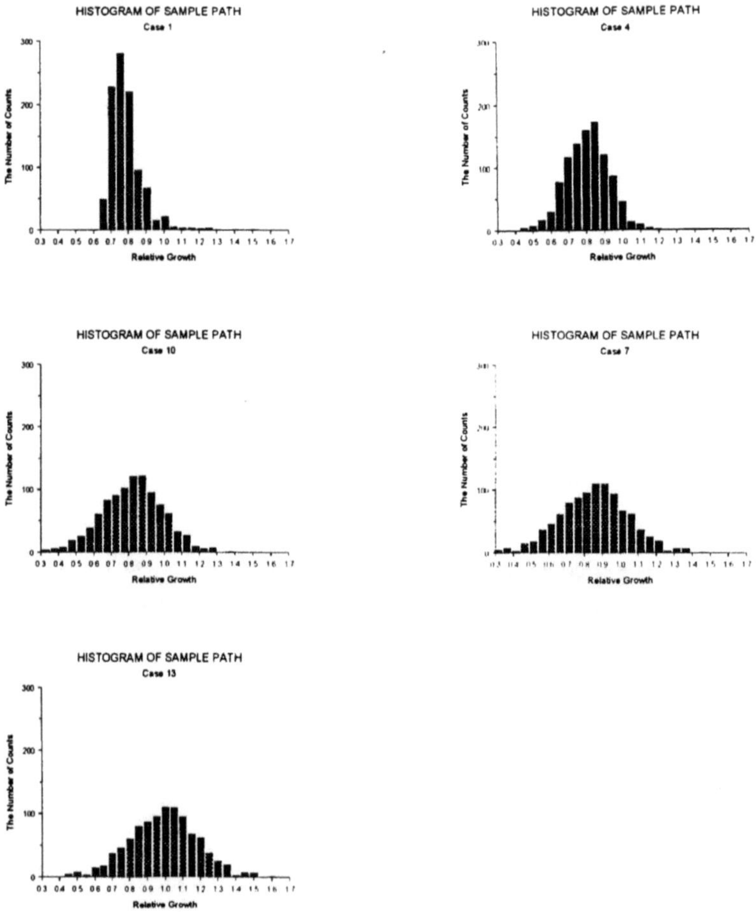

Figure 15. Histogram of the simulated sample path

Even with a small difference in AIC, different stochastic models would lead to different forecasting results. Figure 15 depicts the histogram of a simulated sample path after 10 years of simulation. The number of simulations was 1000, and time increment, Δt, was set to 1/12 year or 1 month. Among thirteen stochastic models in Table 3, four (Case 4, 7, 10 and 13) with the highest AIC and one (Case 1) with the lowest AIC were utilized to show the difference in the histogram. Parameters of each stochastic model was estimated for Karamatsu (*Larix leptolepis*}). Karamatsu showed the largest deviation in AIC over different models.

Case #	α	β	γ	σ	AIC
Case 1	0.9284	-1.2620	4.1233	0.0721	-1808.57
Case 4	0.8717	1.1880	0	0.0374	-1644.40
Case 7	-0.1388	0	0	0.0373	-1645.18
Case 10	0	-0.1849	0	0.0372	-1645.52
Case 13	0	0	0	0.0373	-1646.18

Table 4. Estimated parameters for Karamatsu log price

Estimated parameters for each model were given in Table 4. In order to conduct simulation, the current level of the log price index was set to 1. In the case of the Case 1 model, $dx_t = (\alpha + \beta x_t)dt + x_t^{\gamma}\sigma dB_t$, a mean of the simulated value at year 10 was 0.78954 lower than 1 with a variance equal to 0.0116. A mean and a variance of the simulated value derived from the other models were (0.8057, 0.0141), (0.8529, 0.0363), (0.8223, 0.0301) and (0.9907, 0.0364) for Case 4, 7, 10and 13, respectively. As can be seen in Figure 15, the histogram derived from the Case 1 model was left-skewed, while the others were slightly right-skewed.

CONCLUSIONS

The purpose of this paper was to present stochastic models for timber supply analysis as well as forest stand management. The first stochastic model regarded harvesting events as a stochastic event, then applied a stochastic process to describe forest owner's harvesting behavior. The model utilized a concept of counting process and waiting time process. Although the proposed model is based upon rather simple assumptions, it can be used for forest policy analysis in order to predict how much of percentage of forest owners remain managing forests under certain cost conditions. Recently business environments for forest management in Japan are becoming unfavorable to Japanese local forest owners. This stimulates the owners not to keep forests under management. With use of the proposed stochastic model, it is possible to analyze forest policy to deal with encouraging them to manage forests or not to manage forests. A main advantage of the proposed stochastic model for timber supply is that a nonstationary function was utilized. It is preferable to simulate an economic factor, which was assumed to affect the owner's harvesting decision. Computational experiments conducted here showed that the more the threshold, the longer the forest owners wait for harvesting. Thus the resultant felling probabilities yield the longer harvesting time horizon. Further research has to be conducted to validate the proposed stochastic model with data on harvest events and forest growth.

The second stochastic model dealt with an optimal decision on forest stand management. Stochasticity was only applied to log price dynamics. Stochastic log prices were modeled by the continuous time stochastic process (the geometric Brownian motion), and the binomial option pricing approximation for searching for an optimal rotation age has been embedded into the two-state and one-stage stochastic dynamic programming model. In the proposed stochastic dynamic programming model, three managerial options, {Wait}, {Harvest}-{Keep}, and {Harvest}-{Abandon}, were assumed. In comparison to the deterministic Faustmann approach, our experiments showed that the optimal rotation age under stochastic log prices deviated from that of the Faustmann approach, and showed longer rotation age as the current log price decreased within the range where the {Harvest}-{Keep} decision was implemented. Once the optimal decision became {Harvest}-{Abandon}, the optimal age to abandon decreased while the optimal rotation age from the Faustmann approach increased. This resulted from considering the value of the alternative use, i.e., abandoning the management. With this option, further debt can be hedged. When the current log price is high enough to cover all costs, the optimal rotation age from the stochastic model and the Faustmann approach coincides. However, when the price decreases, they differ greatly. With a downward trend on price dynamics, an optimal rotation age tends to decrease. This may result from depreciation effect on future value. In other words, a downward trend increases the probability that the future price decreases, so that an expected return from future harvesting is depreciated much more than that without such a trend. This stimulates harvesting at an early period with shorter rotation age.

When utilizing stochastic models, estimation problems always remain. We proposed thirteen continuous time stochastic models based on the state dependent volatility process for log prices in order to search for the most appropriate stochastic model for the log prices. Since the log price affects the forest manager's decision most, any error in model specification could result in a non-optimal management decision under uncertainty. Based on our experiments it was shown that the general form of SDVP, Case 1 yielded the best AIC performance for most logs, followed by Case 11but for some logs, other models like in Case 3, 6 and 7 performed best. This implies that the ordinary used Brownian or geometric Brownian motion may not be sufficient to represent the log price stochastic movement. Rather it might be necessary to do comparative analysis of several models before specifying a final stochastic model. We also showed that the local linearization method for parameter estimation was the promising technique to expand our model specification capability, since technical difficulty of parameter estimation has been limiting use of stochastic models other than the Brownian and geometric Brownian model. Using the local linearization method, we believe that a wide variety of the models would be derived to represent a stochastic behavior of log prices. Furthermore we showed that different stochastic models lead to different distribution of simulated sample path at future point in time. For further analysis, it is necessary to investigate how this difference could affect a forest management decision.

LITERATURE

Adams, D.M. and Haynes, R.W. (1980), The 1980 softwood timber assessment market model: Structure, projections, and policy simulations. *For. Sci. Monogr.* 22.

Akaike, H. (1977), 'On entropy maximization principle', in P.R. Krishnaiah (ed.), *Application of Statistics*, North-Holland, Amsterdam.

Amano, M., Noda, I., and Kumazaki, M. (1984). Timber supply projection for private forests in Japan. *In* Proceedings IUFRO Symposium on Forest Management Planning and Managerial Economics. *Edited by* H. Nagumo *et al.*. University of Tokyo, Tokyo. pp.285-292.

Bailey, W. (1991), Valuing agricultural firms: An examination of the contingent-claims approach to pricing real assets, *Journal of Economic Dynamics and Control*, 15:771-791.

Blandon, P. (1985), Forest economics and the Gentan probability distribution. *J. Jpn. For. Soc.* 67: 478-485.

Blandon, P. (1991), Gentan probability and censored sample theory (I). *J. Jpn. For. Soc.* 73: 187-196.

Blandon, P. (1993), Unmanaged forests and forecasting using the Gentan probability distribution. *J. Jpn. For. Soc.* 75: 484-492.

Bowes, M.D. and Krutilla, J.V. (1985), 'Multiple use management of public forestlands', in A.V. Kneese and J.L. Sweeney (eds.), *Handbook of Natural Resource and Energy Economics Vol II*, North-Holland, Amsterdam.

Brennan, M.J. and Schwartz, E.S. (1985), Evaluating natural resource investments, *Journal of Business*,. 58:135-157.

Clark, C.W. (1990), *Mathematical Bioeconomics*, John Wiley & Sons, Inc. New York.

Clarke, H.R. and Reed, W.J. (1989), The tree-cutting problem in a stochastic environment. *J. Econ. Dynamics Control.* 13:569-595.

Cox, J.C., Ross, S.A. and Rubinstein, M. (1979), Option pricing: A simplified approach. *Journal of Financial Economics*, 7: 229-263.

Faustmann, M. (1849), 'On the determination of the value which forest land and immature stands possess for forestry', in M. Gane (ed.), *Martin Faustmann and the Evolution of Discounted Cash Flow*, Institute Paper of the Commonwealth Forestry Institute, 42.

Gilless, J.K. and Buongiorno, J. (1987), PAPYRUS: A model of the North American pulp and paper industry. *For. Sci. Monogr.* 28.

Haight, R.G. and Holmes, T.P. (1991), Stochastic price models and optimal tree cutting results for loblolly pine. *Nat. Res. Model.* 5:423-443.

Morck, R., Schwartz, E., and Stangeland, D. (1989), The valuation of forestry resources under stochastic prices and investments, *Journal of Financial and Quantitative Analysis*, 24:473-487.

Oksendal, B. (1995), *Stochastic Differential Equations*, Springer, New York.

Ozaki, T. (1985), Statistical identification of storage models with application to stochastic hydrology, *Water Resources Bulletin* 21: 663-675.

Plantinga, A.J. (1998), The optimal timber rotation: An option value approach. *For. Sci.*, 44:192-202

Reed, W.J. and Clarke, H.R. (1990), Harvest decisions and asset valuations for biological resources exhibiting size-dependent growth. *Int. Econ. Rev.* 31:147-169.

Reed, W.J. and Ye, J.J. (1994), The role of stochastic monotonicity in the decision to conserve or harvest old-growth forest. *Nat. Res. Model.* 8:47-79.

Richards, F.J. (1958), A flexible growth function to empirical use, *Journal of Experimental Botany*, 10:290-300.

Samuelson, P.A. (1976), Economics of forestry in an evolving society, *Economic Inquiry* 14: 466-492.

Shoji, I. and Ozaki, T. (1997), Comparative study of estimation methods for continuous time stochastic processes, *J. of Time Series Analysis* 18:485-506.

Suzuki, T. (1972), Applications of stochastic process in forestry (I). *J. Jpn. For. Soc.*, 54:234-243. (in Japanese)

Suzuki, T. (1984), The Gentan probability, a model for the improvement of the normal forest concept and of forest planning. *In* Proceedings IUFRO Symposium on Forest Management Planning and Managerial Economics. *Edited by* H. Nagumo *et al.*. University of Tokyo, Tokyo. pp.12-24.

The Science and Technology Agency, (1961), Mokuzaino seisanyosoku ni tuite (Report on forecast of the timber production). Kagakugijutucho-Shigenkyoku Shiryo No.45, The Science and Technology Agency, Tokyo. (in Japanese)*

The Science and Technology Agency, (1963), Mokuzaino seisanyosoku ni tuite (II) -Ringyoniokeru shukakuyoteino sugakutekikenkyu- (Report on forecast of the timber production (II) - Mathematical study on harvest scheduling in forestry). Kagakugijutucho-Shigenkyoku Shiryo No.53, The Science and Technology Agency, Tokyo. (in Japanese)*

Thomson, T.A. (1992), Optimal forest rotation when stumpage prices follow a diffusion process, *Land Economics*, 68:329-342.

Thorsen, B.J. (1999), Afforestration as a real option: Some policy implications. *For. Sci.*, 45:171-178.

Yoshimoto, A. and Shoji, I. (1998), Searching for an optimal rotation age for forest stand management under stochastic log prices, *European J. of Operational Research* 105:100-112.

Zinkhan, F. C. (1991), Option pricing and timberland s land-use conversion option. *Land Economics*, 67:317-325.

*These English titles are tentative translations by the author of this paper from the original Japanese.

Section 3

Economics and Policy

INCORPORATING RISK INTO FINANCIAL ANALYSIS OF FOREST MANAGEMENT INVESTMENTS

W. David Klemperer[1]

ABSTRACT

In calculating net present values of assets or enterprises with risky cash flows, the most common approach is to discount future cash flows with a risk-adjusted discount rate exceeding the risk-free rate. Because decision makers usually do not have the necessary data or expertise to apply more sophisticated approaches to risk analysis, the approach here is to examine refinements to applying risk-adjusted discount rates. With certainty-equivalent analysis, it is shown that, for any given revenue risk, the required addition to the risk-free discount rate becomes less as payoff periods lengthen. Thus it is likely that discount rates commonly applied to long-term forestry investments are too high. Other discussions include present value probability histograms, simulation, problems with the weighted average cost of capital and the capital asset pricing model, and special considerations in discounting risky costs.

Key words: *Risk, finance, forest management, risk-adjusted discount rate, net present value, expected value, certainty-equivalents*

1 Professor of Forest Economics, Virginia Polytechnic Institute and State University, Department of Forestry, Blacksburg, Virginia, USA

K. von Gadow (ed.), Risk Analysis in Forest Management, 149–170.

Introduction

In evaluating and comparing investments, analysts often calculate an investor's maximum bid price or "willingness to pay" for assets or financial ventures. In forestry, such ventures might be purchases of forested properties or, on existing lands, investments in projects like afforestation, pre-commercial thinning, and fertilization. The standard method of finding willingness to pay projects future costs and revenues and calculates "net present value" (NPV), which is the present value of future revenues minus the present value of future costs:

$$NPV = R_o + \frac{R_1}{(1+r)^1} + \frac{R_2}{(1+r)^2} + \frac{R_3}{(1+r)^3} + \ldots + \frac{R_n}{(1+r)^n}$$
$$- C_0 - \frac{C_1}{(1+r)^1} - \frac{C_2}{(1+r)^2} - \frac{C_3}{(1+r)^3} - \ldots - \frac{C_n}{(1+r)^n} \qquad (1)$$

where R and C are revenues and costs in the subscripted years, and r is the investor's desired rate of return. *In equations, all rates, for example r above, will be percent divided by 100.* This chapter's notation is defined in Table 1. An NPV can be viewed as an investor's maximum willingness to pay for an investment, regardless of its actual cost. If an investor paid exactly the calculated NPV, and the asset yielded precisely the projected revenues and costs, the rate of return on the investment would be the interest rate (r) used in calculating the NPV. Investors paying more (less) than the calculated NPV would earn a rate of return lower (higher) than the interest rate used in the calculation. Whether or not an investor uses equation (1), it will be considered here as a valid model for expressing asset values. However, because values for C and R are not known for sure, questions arise in estimating the values for all variables in the equation, including the discount rates for revenues and costs.

CE	=	certainty-equivalent of a risk revenue. CE is the dollar amount which, if received with certainty, would give the investor the same satisfaction or "utility" as the risky revenue with expect revenue, E(R).
E(PV)	=	expected present value
E(R)	=	expected value of a risky revenue at one time such as that shown in Figure 1.
k	=	risk premium in the RADR (percent/100). $r_a = r_f + k$
n	=	payoff period or number of years between initial investment and final return.
NPV	=	net present value, or present value of revenues minus present value of costs.
PV_{CE}	=	certainty-equivalent present value of a risky future revenue (i.e., a future CE discounted with the risk-free discount rate). This is considered the correct present value.
r_f	=	real risk-free discount rate (percent/100), assumed here to be .03.
r_a	=	risk-adjusted discount rate for discounting single expected values (percent/100).
R_i	=	revenue in year i
C_i	=	cost in year i

Table 1. Notation

If the values of R and C are known with certainty, for example, contractually agreed upon revenues like bond yields or costs of predetermined loan payments, we have a case of risk-free cash flows. In that case, in the above equation, it is appropriate to use a risk-free discount rate reflecting market returns on risk-free investments. However, it is more common for future cash flows to be risky, in other words, exact values are not known with certainty. Examples are common in forestry: we know that if we plant the same tree species repeatedly on similar sites, future harvest revenue can vary widely due to mortality, planting stock, weather, insects, disease, fire, stumpage price fluctuations, and other factors. Or if we set aside an income-generating recreation area, the value of entry fees over time could vary, depending on population trends, income levels, people's preferences, weather, and substitute recreation sites. In such cases, projecting a single revenue for a given event at a set time implies an average of possible outcomes but says nothing about the amount of variation, which is called *risk*. The positive side of revenue risk is the possibility of earning more than the average return, and the negative is the probability of earning less. On average, for most investors, the disadvantage of risk outweighs the advantage; we call such investors "risk-averse".

This chapter's objective is to explain the nature of financial risk and to outline practical ways of incorporating risk into financial calculations in forestry investments. Emphasis will be less on cutting edge theory than on applicability in actual analyses. Unless otherwise noted, assume that all values and discount rates are real without inflation, before taxes. Initially only revenues will be considered.

Risky vs. Risk-Free Revenues

Figure 1 shows possible yields from a hypothetical risky investment after a 10 year period. This bar chart is a *probability histogram* showing the probabilities of receiving various incomes. It shows that the probability is .2 that revenue will be between $0 and $8,000, or, on average, there is a 20 percent chance of receiving the midpoint value of $4,000. Following this reasoning, if you made this investment hundreds of times, 20 percent of the time it would yield $4,000, 60 percent of the time $12,000, and 20 percent of the time $20,000. The probabilities must sum to 1 or 100 percent, and many repetitions of the investment would yield an *expected value* for the revenue of $E(R) = .2(4{,}000) + .6(12{,}000) + .2(20{,}000) = \$12{,}000$. We also call this an *expected revenue*. The expected value of any set of revenues is the sum of the possible revenues multiplied by their probabilities of occurrence.

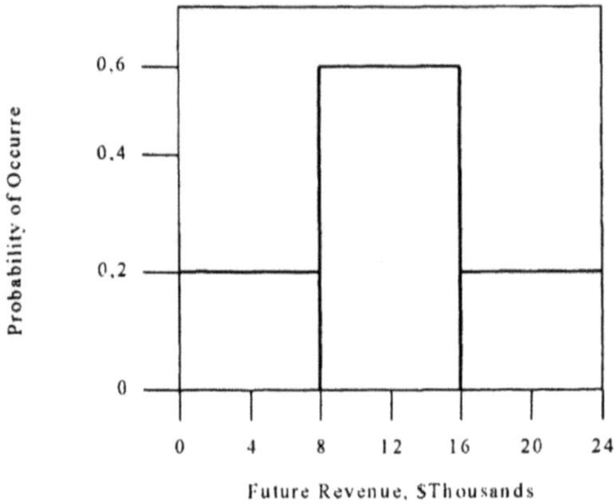

Figure 1. Risky Revenue with a $12,000 Expected Value in 10 Years

One problem is that an expected value does not reflect its variance. For example, the Figure 1 revenue risk would be greater if income could be expected to vary from, say, -$12,000 to +$36,000, and still average $12,000. The greater the possible variation in a cash flow, the greater its risk. One often uses statistical measures like the variance, standard deviation or coefficient of variation to measure risk in an expected value. Note that Figure 1 is a symmetrical probability distribution where the expected value equals the mean of (4,000 + 12,000 + 20,000)/3 = $12,000. Sometimes distributions are skewed so that the expected value is greater than or less than the mean.

Analyses here will assume that risky outcome probabilities can be estimated, either subjectively or statistically. *Uncertainty* refers to situations where one cannot estimate the probability of an outcome's occurrence. For example, we do not know the probability that a recession will occur next year or that a war will cause certain price shocks. *Risk* indicates that, based on past experience, we can estimate the probability of some outcome like a fire during a given time. This chapter deals mainly with risk, although uncertainty is mentioned under "Simulation".

Now consider a risk-free income with no variation. An example is a $12,000 guaranteed yield from a 10-year government bond pictured in Figure 2. Assume an inflation-indexed bond without the risk of losing value due to unanticipated inflation. Thus, the government will guarantee this yield with 100 percent certainty to the purchaser (probability = 1). The expected value is $12,000 (or (1)12,000) - the same as the E(R) of the Figure 1 revenue, except that the bond yield has no variation or risk.

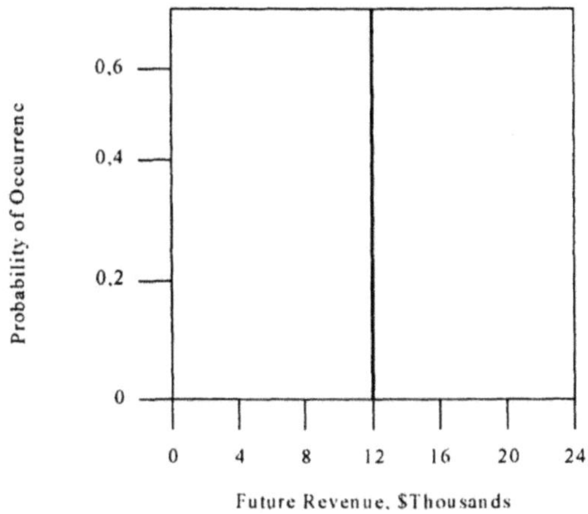

Figure 2. Risk-Free Revenue with a $12,000 Expected Value in 10 Years

Now ask, "Which $10,000 expected revenue is worth more to me, that in Figure 1 or 2?" Most investors would answer, "Figure 2," and are therefore considered *risk-averse*; given the same expected value, they prefer lower variation in returns. Those indifferent between the Figure 1 and 2 revenues are called *risk-neutral*; as long as the expected value is the same, the amount of variation in returns makes no difference to them. Those desiring greater variation are *risk seekers*; for them, the pleasure from the chance of higher-than-average returns more than offsets the displeasure from possibly receiving less-than-average returns.

Financial Analysis of Risk-free and Risky Revenues

Assets with no risk earn risk-free rates of return. Their present values should therefore be calculated with a risk-free discount rate. Risk-free rates of return in the United States have historically averaged about 1-3 percent. Simon (1990) projects a long term real risk-free rate of return, r_f, of 3 percent, the value to be used here. Recall that this excludes inflation; if annual inflation averaged 4 percent, the actual or "nominal" risk-free interest rate would be 100[(1.03)(1.04) - 1] = 7.12 percent. Assuming investors with a real 3 percent risk-free required rate of return, the correct present value of the risk-free revenue in Figure 2 is:

$$\textit{Correct present value of risk-free revenue} = \frac{\textit{risk-free revenue}}{(1+r_f)^n} = \frac{12{,}000}{(1.03)^{10}} = \$8{,}929 \quad (2)$$

Thus, assuming average investors desiring real 3 percent risk-free rates of rates of return, the bond would tend to sell for $8,929 today if it were to yield $12,000 in 10 years (constant dollars). That would guarantee buyers a sure 3 percent real rate of return. Anyone requiring a risk-free rate of return greater than 3 percent would pay less than $8,929 for the bond.

When evaluating a future expected revenue, most investors, being risk-averse, assign a lower present value to a return with greater variance. Consider the risky and risk-free revenues in Figures 1 and 2, both yielding the same $12,000 expected value in 10 years. Equation (2) shows that the correct present value of the Figure 2, risk-free $12,000 is $8,929 for an investor with a 3 percent risk-free discount rate. If risk-averse, that same investor would discount the Figure 1 risky expected revenue of $12,000 with a *risk-adjusted discount rate* (r_a) higher than the risk-free rate to yield a lower present value reflecting the lower satisfaction due to risk. Thus, with risk aversion,

$$\begin{gathered}\textit{Present value of a given expected revenue} = \\ \frac{\textit{Risky } E(R)}{(1+r_a)^n} < \frac{\textit{Risk-free } E(R)}{(1+r_f)^n} \\ \textit{or for our example,} \\ \frac{\textit{Risky expected } \$12{,}000}{(1+r_a)^{10}} < \frac{\textit{Risk-free } \$12{,}000}{(1.03)^{10}}\end{gathered} \tag{3}$$

Since the numerators are the same, the risk-adjusted discount rate must exceed the risk-free discount rate for the inequality to hold.

Financial markets reveal that investors who are satisfied with, say, a 3 percent real return on risk-free bonds demand higher risk-adjusted rates of return on riskier investments. This demonstrates that, on average, investors are indeed risk averse. Thus, the remaining analyses will assume risk aversion, unless otherwise stated. The difficult question is to determine how much above the risk-free discount rate the risk-adjusted rate should be.

Risk-adjusted Discount Rates

The U.S. Office of Management and Budget (OMB) recommends that risky expected cash flows and other estimated values from government projects be evaluated with a 7 percent real, risk-adjusted discount rate (USOMB 1992). This reflects an average before-tax rate of return on private capital in the U.S. at average risk levels. However, not all agencies follow the guide (the U.S. Forest Service uses a 4 percent real discount rate). The OMB reasons that public and private projects should earn the same rates of return. If that does not occur, OMB supporters argue that you could shift capital from lower to higher rates of return and society would gain (Hirshleifer 1964)[2]. One problem with using a single risk-adjusted discount rate like OMB's 7 percent is that it applies to investments of average risk and average duration. For any given size investment of a given duration, to provide the same satisfaction to a risk-averse investor, low risk investments need not earn as much as 7 percent, while some high risk ones must earn more. Thus, OMB's fixed 7 percent required earning rate favors high risk ventures and works against low risk investments.

How can one find the appropriate risk-adjusted discount rate for discounting a risky expected revenue like the $12,000 in Figure 1? We could ask the investor, "what sure income - the *certainty-equivalent* - would give you the same satisfaction as the risky revenue?" To reflect the reduced value due to risk, a risk-averse investor would assign a certainty-equivalent lower than the expected revenue (CE < E(R)). A future certainty-equivalent, being implicitly risk-free, should be discounted with the risk-free rate r_f, resulting in the correct present value of a risky expected revenue: PV_{CE}. One could also get the same correct present value by discounting the risky expected revenue (E(R)) with an $r_a > r_f$. The appropriate risk-adjusted discount rate r_a results in the following equality (Robichek and Myers1966):

$$\textit{Correct present value} = \frac{CE}{(1+r_f)^n} = PV_{CE} = \frac{E(R)}{(1+r_a)^n} \qquad (4)$$

Thus, an expected revenue's certainty-equivalent discounted with a risk-free rate should equal the risky expected revenue discounted with a risk-adjusted rate. Under risk aversion, E(R)>CE, and therefore $r_a > r_f$. Taking the last 2 terms of equation (4) and solving for the risk-adjusted discount rate r_a:

$$(1+r_a)^n = \frac{E(R)}{PV_{CE}}$$
$$r_a = \sqrt[n]{\frac{E(R)}{PV_{CE}}} - 1 \qquad (5)$$

[2] The opposing argument is that the large numbers of government projects and taxpayers provide a risk pooling effect where unusual gains and losses offset one another in any year. That view suggests a risk-free discount rate for risky government projects (Samuelson 1964). But the fact remains that decision makers in government agencies often behave like risk-averters. Arrow and Lind (1970) suggest that if society's risk-free time preference rate is, say, 3 percent, the private sector may be under-investing at, say, 7 percent (avoiding the lower rate of return projects) because firms and individuals cannot take advantage of risk pooling, which society can do collectively. This could be seen as a market failure. Although the issue of a government risk premium is unresolved, the analysis here does not distinguish between private and public risk premiums.

Suppose your certainty-equivalent of the \$12,000 risky expected revenue in Figure 1 was \$8,000. Substituting these values into equation (4), and remembering that n=10, and r_f=.03, you can solve for your risk-adjusted discount rate r_a:

$$Correct\ present\ value = \frac{8{,}000}{(1.03)^{10}} = \$5{,}953 = \frac{12{,}000}{(1+r_a)^{10}}.$$

$$Solving\ for\ r_a\ as\ in\ equation\ (5), \qquad (6)$$

$$r_a = \sqrt[10]{\frac{12{,}000}{5{,}953}} - 1 = .07262,\ or\ about\ 7.3\%$$

As a check, discounting the risky \$12,000 at 7.262 percent for 10 years will give the correct present value of \$5,953.

The amount by which the risk-adjusted discount rate exceeds r_f is the *risk premium*:[3]

$$Risk\ premium = k = r_a - r_f \qquad (7)$$

In the above example, the risk premium is 7.3 - 3 = 4.3 percentage points. Unfortunately, no universally correct risk premium exists; it depends on the payoff period, the amount of risk in the revenue, and the investor's degree of risk aversion, as seen below.

Variation in Risk-adjusted Discount Rates

Under even-aged softwood management, rotation lengths vary from 20-35 years in the southern U.S, 40-80 years in the Pacific Northwest, and can exceed 80 years if one considers non-timber values. For some warmer regions, fiber crop rotations can be 5 years or less. For different payoff periods, how great would the perceived risk need to be for OMB's 7 percent risk-adjusted discount rate to be correct? You can answer this by solving equation (4) for the certainty-equivalent:

$$Correct\ present\ value = \frac{CE}{(1+r_f)^n} = \frac{E(R)}{(1+r_a)^n}$$

$$Solving\ for\ CE, \qquad (8)$$

$$CE = \frac{E(R)(1+r_f)^n}{(1+r_a)^n}$$

[3] Remember that here we are dealing only with risky *revenues*. Later it is shown that under some conditions, the risk premium for risky *costs* is *negative*, in which case the risk adjusted discount rate would be less than the risk-free rate.

This says that if the risk-free rate is r_f and one discounts a risky expected revenue at a risk-adjusted rate of r_a, for an n-year payoff period, the implied certainty equivalent would need to be CE for the present value to be correct. For example, using equation (8), assuming a \$10,000 risky revenue, E(R), received after different payoff periods, the last column of Table 2 shows future certainty-equivalents implied by using OMB's 7 percent discount rate, if 3 percent is the risk free rate, which means a 4 percentage-point risk premium. In the first row, under these conditions, the certainty-equivalent of a \$10,000 risky income expected from, say, a stock sale in 5 years would need to be \$8,265, using equation (8):

$$CE = \frac{10{,}000(1.03)^5}{(1.07)^5} = \$8{,}265$$

This means that, if a 7 percent risk-adjusted discount rate is correct, an investor must be indifferent between a \$10,000 risky return in 5 years and a sure \$8,265 received at the same date. This is a reasonable result which says that a sure \$8,265 in 5 years discounted at the risk-free 3 percent gives the same present value as the risky \$10,000 discounted for 5 years at the risk-adjusted 7 percent—about \$7,130 in either case. The second row of Table 2 is likewise reasonable: One could easily be indifferent between a risky \$10,000 Christmas tree revenue expected in 10 years and the calculated sure \$6,832 shown in the last column. Thus, with such payoff periods, OMB's 7 percent discount rate could be reasonable for many investors.

Payoff period (n, years)	Risky future expected revenue (E(R), dollars)	Required future certainty equivalent (CE, dollars)*
5	\$10,000 (e.g., stock market)	\$8,265
10	\$10,000 (e.g., Christmas trees)	\$6,832
30	\$10,000 (e.g., Loblolly pine)	\$3,189
50	\$10,000 (e.g., Douglas-fir)	\$1,488
100	\$10,000 (e.g., long rotation hardwood)	\$221.5

*Equation (8). The required certainty-equivalent (CE) is the dollar amount which, if received with certainty at the end of the payoff period, would have to give the investor the same satisfaction as the risky expected revenue of \$10,000 in column 2, if 7 percent were the correct real risk-adjusted discount rate. Column 2 discounted with a 7 percent risk-adjusted discount rate gives the same present value as column 3 discounted at the 3 percent risk-free rate. If required CE's at longer payoff periods seem too low, this would suggest that the correct risk-adjusted discount rate should be less than the 7 percent used here

Table 2. Required Future Certainty Equivalents for a 7 percent Risk-Adjusted Discount Rate to Give Correct Present Value (r_f=.03)

Now consider the longer payoff period of 30 years in row 3 of Table 2. It might seem questionable for some investors to be indifferent between the risky $10,000 loblolly pine expected revenue in 30 years and the $3,189 computed certainty-equivalent on the same date. Rows 4 and 5 of Table 2 show that for a 7 percent risk-adjusted discount rate to be appropriate, an investor would have to be indifferent between an expected $10,000 Douglas-fir harvest in 50 years and a guaranteed $1,488 on the same date... or a public agency with a long time horizon would be indifferent between an expected $10,000 hardwood yield in 100 years and a guaranteed $221.5 on the same date. (Discounting $221.5 at a 3 percent risk-free rate for 100 years gives the same present value as the risky $10,000 discounted at 10 percent -- about $11.50 in either case.)

Often private firms use real discount rates of 12 percent or higher, after taxes, to evaluate investments. Thus, required rates of return would be higher before taxes. Table 3 is analogous to Table 2, except that the risk adjusted discount rate is raised from 7 to 12 percent. In that case, the implied certainty equivalents of a risky $10,000 range from $6,578 for a 5 year payoff period to $2.30 for a 100 year payoff period. Again, the required certainty equivalents seem reasonable for shorter payoff periods but appear to be increasingly lower-than-expected as payoff periods lengthen.

Payoff period (n, years)	Risky future expected revenue (E(R), dollars)	Required future certainty equivalent (CE, dollars)*
5	$10,000 (e.g., stock market)	$6,578
10	$10,000 (e.g., Christmas trees)	$4,327
30	$10,000 (e.g., Loblolly pine)	$810
50	$10,000 (e.g., Douglas-fir)	$1528
100	$10,000 (e.g., long rotation hardwood)	$2.30

*Equation (8). The required certainty-equivalent (CE) is the dollar amount which, if received with certainty at the end of the payoff period, would have to give the investor the same satisfaction as the risky expected revenue of $10,000 in column 2, if 12 percent were the correct real risk-adjusted discount rate. Column 2 discounted with a 12 percent risk-adjusted discount rate gives the same present value as column 3 discounted at the 3 percent risk-free rate. If required CE's at longer payoff periods seem too low, this would suggest that the correct risk-adjusted discount rate should be less than the 12 percent used here

Table 3. Required Future Certainty Equivalents for a 12 percent Risk-Adjusted Discount Rate to Give Correct Present Value (r_f=.03)

There is little evidence that timber investors think most long term forestry ventures are so risky that certainty-equivalents would be as low as Tables 2 and 3 indicate for payoff periods longer than 30 years. To get higher certainty-equivalents which reflect less risk for the longer payoff periods, one would substitute a risk-adjusted discount rate less than 7 percent or 12 percent in equation (8). Numerous sources suggest that forestry in most regions of the U.S. is not viewed as a high risk venture.[4] In addition, forest asset values often vary less than common stock portfolios, and are even less risky in a diversified portfolio, because correlation of forest values with common stock indexes is often zero or negative. Trees have the added advantage that, unlike agricultural crops, they store well unharvested. And if stumpage prices plummet, one can postpone harvest and cash in on later price spikes. Thus, price variability can increase present values of forests (Haight 1991). Although a good deal more research is needed, Tables 2 and 3 suggest that appropriate risk-adjusted discount rates in forestry may often decline with increasing payoff period. Similar conclusions have been reached for current valuation of future income from stocks (Chen 1967).

Note that we should not generalize timber investment conclusions from Tables 2 and 3 to all investments. The above inferences apply only to most timber investments in the United States or countries with similar economies and climates. Very risky investments, say, building factories in politically unstable countries, would have much larger revenue variance than many timber investments and could thus have relatively low certainty equivalent values for future income. Therefore, regardless of payoff periods, one would generally discount expected yields from such high-risk investments with high interest rates.

If investors make optimistic projections of timber harvest revenues, ignoring the possibility of less desirable outcomes, income variability is underestimated and such revenues will exceed true expected values. Optimistic revenues could be justifiably discounted with rates higher than the risk-adjusted rates discussed here, which are meant only for discounting expected revenues. To avoid overly-optimistic forest income projections, investors should make some effort to estimate subjective probabilities of a few possible outcomes and calculate expected values of each scenario. For example, a forest manager may project a 25 percent chance of receiving harvest yields in 30 years averaging either \$1,000, \$3,000, \$5,000, or \$7,000, yielding an expected revenue of:

$$.25(1{,}000) + .25(3{,}000) + .25(5{,}000) + .25(7{,}000) = \$4{,}000.$$

[4] See Mills and Hoover 1982, Fortson 1986, Webb *et al.* 1987, Redmond and Cubbage 1988, Conroy and Miles 1989, Zinkhan *et al.* 1992.

Tables 2 and 3 suggest that appropriate before-tax risk-adjusted discount rates for many long-term forestry investments may be well below OMB's 7 percent. The latter was based largely on studies which divide firms' annual before-tax income by the market value of assets, thus averaging all long- and short-term projects together.[5] That approach yields a risk premium which could often be too high for many long-term forestry ventures.

If we knew certainty-equivalents of revenues, we could simply discount CEs with a risk-free rate, and risk-adjusted discount rates would not be needed. Without such information, public and private analysts continue to use risk-adjusted discount rates in investment analysis. However, we cannot prescribe a universal forestry discount rate because correct risk-adjusted rates depend on probability distributions of cash flows, payoff periods, and investors' degrees of risk aversion (for a more detailed analysis of these relationships, see Klemperer *et al.* 1994).

Efficient Capital Allocation

Inefficient resource use would result if all projects were required to earn some fixed average industrial rate of return, say OMB's 7 percent. Assuming a 3 percent risk-free interest rate, suppose that the approach taken in Tables 2 and 3 yielded 7 percent as reasonable risk-adjusted discount rate for shorter term industrial investments (a risk premium of 7 - 3 = 4). Suppose, on average, the correct risk-adjusted discount rate for long term forestry projects were 5 percent (a 2 percent risk premium), but all projects, including forestry were required to earn 7 percent. That would imply that the risk-free earning rate in forestry would be 5 percent (7 minus forestry's 2 percent premium). This implied risk-free return should be the relevant rate for comparing performance of investments with different risks and payoff periods. If the economy-wide average risk-free earning rate on short-term projects is 3 percent (4 percent risk premium), this would mean under-investment in forestry under the foregoing assumptions. Shifting capital from 3 percent to forestry's 5 percent (risk-free) would increase returns to aggregate investment. An optimal investment allocation would occur when forest output prices declined, due to higher production, to the point where the implied risk-free earning rates on the last dollars invested in forestry would be equal to the economy-wide 3 percent for the last dollars invested in all other ventures. Then no investment reallocation could increase total

[5] See Stockfisch (1969). A later example is Nordhaus (1974).

returns. Such an optimum leaves room for substantial differences in risk-adjusted earning rates.

Thus, it is not inefficient if certain forestry investments are evaluated with, say, a 5 or 6 percent risk-adjusted rate, despite higher returns observed elsewhere. Differences in average rates of return can be acceptable if they represent varying risk levels and payoff periods. *The efficiency guide for capital allocation is that rates of return on added investment should be equal for all enterprises on an implied risk-free basis, not risk-adjusted.*

Present Value Probability Histograms

One can avoid dealing with risk-adjusted discount rates by constructing *present value probability histograms* of future risky cash flows. Consider the risky revenue in Figure 1. Each bar can be seen to occur a given percent of the time with certainty. The graph, for example, shows that, with many repetitions of the investment, in 10 years you would receive \$4,000 twenty percent of the time, or \$12,000 sixty percent of the time, etc. Thus, in a sense, each bar is certain and can be discounted with a risk free discount rate. Discounting for 10 years at the 3 percent risk-free rate gives the present value probability distribution in Figure 3.

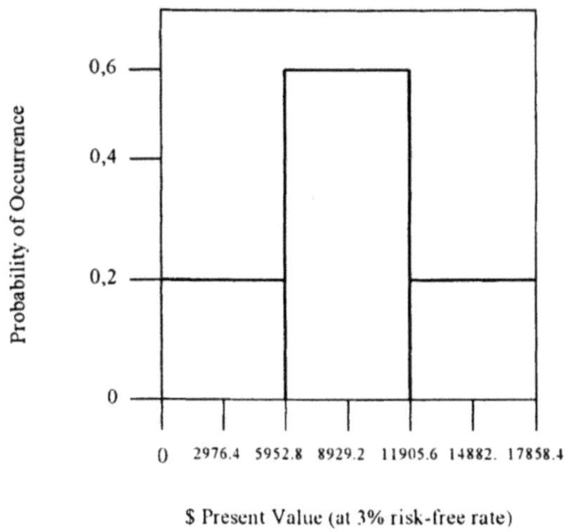

Figure 3. Present Value Probability Histogram of Figure 1 Risky Revenue

For example, the first midpoint value is the corresponding value from Figure 1 discounted to the present: $4{,}000/(1.03)^{10} = \$2{,}976$. To get the expected present value for Figure 3, sum the products of the mid-point values of the bars and their probabilities of occurrence:

$$E(PV) = 0.20(2{,}976) + 0.60(8{,}929) + 0.20(14{,}882) = \$8{,}929$$

For a risk-averse investor, the $8,929 exceeds the correct present value because E(PV) is risky (has variation).

Given probability distributions for positive and negative cash flows of an investment, a computer program can simulate its net present value (NPV) probability histogram, using a risk free discount rate. For several projects, a decision maker could subjectively evaluate NPV histograms and their associated expected present values. Although NPV distributions are an excellent way to compare present values of risky alternatives, they usually require complex computer programs which might not be readily available to many investors. Most decision makers still prefer the simplicity of a single present value obtained with a risk-adjusted discount rate. Thus, we are still left with the problem of determining such rates.

Simulation

Computer simulation can generate net present value probability histograms of the type shown in Figure 3. But to do so requires probability distributions for all a project's cash flows for a reforestation project. These could be planting cost, annual costs and revenues, thinning costs, thinning revenues, fertilization costs, and final harvest revenue. The computer program can simulate the investment hundreds of times, each time calculating and saving the net present value computed with a risk free discount rate. With each simulation, the program randomly selects a value for each of the cash flows according to its distribution. For example, if Figure 1 was a harvest value distribution, 20 percent of the time the program would select a value from the $4,000 range, 60 percent of the time from the $12,000 range, and so on. After computing hundreds of NPVs in this way, the program can construct an NPV probability distribution based on the number of times each present value is generated. This procedure is called *Monte Carlo simulation*.

Instead of computing NPVs, simulation programs can calculate hundreds of internal rates of return (IRR) for a risky project and then construct its IRR probability histogram with IRR on the horizontal axis and probability of occurrence on the vertical axis. Managers can subjectively choose between such histograms for projects of similar size and duration.

A method of dealing with investment uncertainty, where outcome probabilities are not known, is to first calculate a profitability criterion like net present value or rate of return for an investment's most likely cash flows. Then one can test how profitability changes as selected variables are set at their low and high levels. Such *sensitivity analysis* often reveals that profitability is more sensitive to some variables than others. Analysts can then concentrate on refining estimates of variables to which returns are most sensitive. This is a way to deal with *uncertainty,* where outcome probabilities are not known.

Risky Costs

Because risk-averse investors reduce the present value of risky revenues, Brown (1983) argued that a risky *cost's* present value should be made *larger* (in absolute value terms) to account for disutility from risk. For a risk-averse investor, the satisfaction from knowing that a risky cost may be less than its expected value is outweighed by the displeasure from the possibility of incurring higher-than-average costs. Therefore, Brown suggested that risky costs should be discounted with risk-adjusted rates *lower* than the risk-free rate, thus making their negative present values greater than would be the case if they were sure. As with risk-free revenues, risk-free costsBfor example, loan paymentsBshould be discounted with a risk-free rate. Future risky costs can therefore be under-estimated by the "standard practice" of discounting revenues and costs with the same risk-adjusted discount rate higher than the risk-free rate.

Certainty-Equivalents Again

For any investment scenario, the theoretically correct optimal management plan would be that which yielded the maximum net present value of the investor's certainty-equivalents discounted with a risk-free discount rate. Recall that the certainty-equivalent of an uncertain expected revenue (cost) is that sure sum which would bring the investor the same utility (disutility) as the expected revenue (cost). For a risky revenue, it is well-known that a risk-averse investor will cite a certainty-equivalent which is lower than the expected value of the risky revenue (Robichek and Myers 1966). Following that reasoning, the same investor's certainty-equivalent of a risky cost ought to *exceed* that cost's expected value. If the correct present value of an expected risky cost is its larger certainty-equivalent discounted with the risk-free interest rate, that same correct present value could thus be obtained by discounting the cost's expected value with a discount rate less than the risk-free rate.

Correlation of Costs and Revenues

Lewellen (1977) noted that costs which have a perfect positive correlation with revenues can simply be offset by revenues in the year they occur. In such cases, net revenues can be discounted with a risk-adjusted discount rate above the risk-free rate, which is the standard practice of discounting costs and revenues with the same risk-adjusted rate.[6] But Celec and Pettway (1979), argue that perfect correlation of costs and revenues (for example income taxes and income) is the exception rather than the rule. They thus recommend that in most cases, risky costs should be discounted with rates lower than those used for revenues. This conclusion is now found in some finance texts (for example, see Brigham and Gapenski 1991). Celec and Pettway's conclusion was a step in the right direction, but did not go as far as the correct conclusion that risky costs uncorrelated with revenues should be discounted at rates below the risk-free rate.

One unresolved issue is how to measure the degree of correlation between risky costs and revenues. In a given period, if revenues offset costs, and both are perfectly correlated, the same discount rate can apply to revenues and costs. As correlation diverges, say by measure of the correlation coefficient, the discount rate for expected costs drops below that for expected revenues. We need to explore the implications of different degrees and measures of correlation between costs and revenues.

When risky costs include possible losses of unique ecosystems, the literature on the "endowment effect" may be relevant (Kahneman et al. 1990, and Shogren *et al.* 1994). This effect refers to people's resistance to giving up non-market goods which have no perfect substitutes. This provides an added rationale decreasing the discount rate for certain types of risky environmental costs in order to boost their present value.

Discounting Risky Costs

In forest investment analysis, the issue of discounting costs raises important questions which need more research. To what extent is incorrect discounting of future costs in forestry leading to inefficient decisions? What are some of the theoretical questions or unresolved issues in the area of discounting costs? How important are these issues in evaluating future damage to forest ecosystems?

As an example, consider the following hypothetical probability distribution for forest damage caused by fires in a region, ten years from now, in constant dollars: the probability of $40 million damage is 20 percent, $80 million is 40 percent, $120 million is 30 percent, and

---

[6] This is Prince's (1985) argument, except that he does not recognize that using the same risk-adjusted discount rate for costs and revenues is only appropriate when the two are perfectly correlated.

\$160 million is 10 percent. The expected value of damages (or expected cost) is .2(40)+.4(80)+.3(120)+.1(160)=\$92 million. Discounting with OMB's 7 percent, the present value of damages is $92/(1.07)^{10}$ = \$46.8 million, which would be today's willingness to pay to avoid the damage. But if we assume damage costs are uncorrelated with forest revenues and discount these risky costs with a reduced rate, say, 2 percent (below the risk-free rate), to reflect the disutility from damage variation, the present value of expected damage is \$75.5 million, or over 60 percent higher. Such valuation discrepancies obviously have enormous policy implications. The cost discounting issue is important in public and private benefit/cost analyses, many of which use a real discount rate of 7 percent or higher for both revenues and costs. Does this result in serious under-estimation of some risky costs?

Given the variable risks involved in resource management costs, the long time horizons in resource planning, the large public and private investments made in natural resources, the potential for environmental damages, and the importance of future costs in most investments, the question of correct discounting of risky costs is extremely important. It behooves us to examine this issue more closely.

Discounting Borrowers' Risk-free Loan Payments

As another example of using different discount rates for costs and revenues, consider the impact of a 10-year, \$100,000 loan on a project's present value. If the real loan interest rate is 5 percent and includes a 4 percent projected inflation rate, the nominal loan interest is (1.05)(1.04)-1=.092, or 9.2 percent. Using a capital recovery formula, annual nominal fixed loan payments are \$15,719.50 per year. Since the payments are fixed in current dollars, calculations must be made with nominal interest rates which include inflation.

Suppose the investor discounts project cash flows with an 8 percent real rate and projects 4 percent inflation, yielding a nominal discount rate of (1.08)(1.04)-1=.1232, or 12.32 percent. Using the typical approach of discounting all cash flows with the same discount rate (e.g. the one above 12.32 percent), the loan's impact on project present value would be:

$$100{,}000\ -\ 15{,}719.50\ \frac{(1.1232)^{10}\ -1}{.1232(1.1232)^{10}}\ =\ 100{,}000\ -\ 87{,}667\ =\ \$12{,}333, \qquad (9)$$

or the loan amount minus the present value of loan payments, using the nominal discount rate, since the loan payments are fixed in nominal terms.

However, the *borrower's* present value of a fixed payment loan should be calculated with a *risk-free* interest rate, separately from present value of the risky cash flows. From the borrower's view, loan cash flows are perfectly certain: the borrower knows that the $100,000 income will occur, and the firm has signed a contract to make payments. If the risk-free real discount rate is 3 percent, the nominal risk-free rate is (1.03)(1.04)-1=.0712, or 7.12 percent, given the 4 percent projected inflation rate. From the borrower's view, the loan adds the following present value to the timberland purchase, computing with the nominal risk-free discount rate:

$$100{,}000 - 15{,}719.50\,\frac{(1.0712)^{10} - 1}{.0712(1.0712)^{10}} = 100{,}000 - 109{,}797 = -\$9{,}797 \tag{10}$$

Since the present value of loan payments now exceeds the loan amount, the net present value of the loan is negative. Because the firm can save on its income taxes by deducting loan interest, the net impact of the loan may not be negative. Another twist is that borrowing money makes the timberland venture more risky to the borrower, since there is now the obligation to pay off the loan, no matter what the income from the tract will be, thus increasing the risk of bankruptcy. That could boost the appropriate discount rate for the tract's other cash flows, and reduce NPV to the buyer.

The literature is rife with disagreement about the effect of debt on a firm's value (for example, see See Modigliani and Miller 1958 and 1953). The key point to remember here is: *When computing the present value of a loan, the borrower should* ***not*** *use a risk-adjusted discount rate higher than the risk-free rate.* For risky revenues, we increase the discount rate above the risk-free rate to decrease present value, thus reflecting the disutility from revenue variance. But there is no reason for the *borrower* to reduce the present value of loan payments by increasing the discount rate as was done in equation (9).[7] This over-states NPV by understating the present value of loan payments. Since the borrower's correct loan present value is minus $9,797, equation (9) over-states the loan NPV by the difference between the two values, or $22,130, ignoring tax impacts. *Such errors can cause over-bidding for properties and over-valuation of leveraged assets in general.* Investors who over-pay for assets will not earn their desired rate of return, adjusted for risk.

[7] It is incorrect to assume that, every year, income will offset the borrower's loan payments and to just discount this net revenue with the risk-adjusted rate exceeding the risk-free rate. That would be like discounting costs and revenues separately with same the rate. But that view ignores the fact that loan payments are completely independent of revenues. Whether income is high, low, zero, or negative, loan payments are always due (unless a borrower declares bankruptcy). Discounting loan payments with a risk-free rate reflects the disutility from this unavoidable certainty of costs.

Let us now turn to considering problems with other means of setting discount rates.

The Weighted Average Cost of Capital

Business analysts sometimes suggest discounting risky cash flows with the „weighted average cost of capital" (WACC), which is an average interest rate a firm pays on borrowed capital. When raising capital for new investment, a corporation can borrow money or issue new shares of stock. Borrowing money requires paying interest, which is called the cost of debt. On stock issued, a firm eventually will pay dividends, which are the cost of equity. Equity refers to the share of ownership which stockholders have in the firm. The cost of debt and equity combined is the cost of capital, which, if averaged and weighted by the firm's percentages of debt and equity, becomes the WACC. As an example of a WACC, suppose 30 percent of a firm's capital were raised by issuing stock on which a 5 percent dividend was paid, and 70 percent were borrowed at 9 percent interest. The WACC would be .30(5)+.70(9)=7.8 percent. For the after-tax cost of capital, firms consider the tax savings from deductibility of interest in calculating income taxes.

Firms sometimes use the WACC as a minimum acceptable rate of return or „hurdle rate" that new projects should earn. But when the cost of capital is below widely available alternative rates of return, it would be inefficient to invest in projects yielding less than available alternatives. Several studies have found business' hurdle rates three or four times the interest cost of borrowing funds. Thus, the WACC is unlikely to represent typical business hurdle rates, especially with risky investments (Dixit 1992). Moreover, the WACC, being fixed for a given firm, cannot adjust for the fact that, for any given variance or risk in a future revenue, the risk premium in the discount rate should decline with lengthening payoff period (see Tables 2 and 3).

Capital Asset Pricing Model

When comparing returns in a portfolio of diversified investments, one measure of a single investment's risk is its "beta", which measures that investment's variability relative to the portfolio or market as a whole. For example, a stock will have a beta of 1 if its value moves up and down by the same percent as the entire market. An asset with a beta of 2 is twice as variable as the market, and a beta of 0.5 suggests less risk, since volatility is half that of the market. According to the capital asset pricing model (CAPM), the higher the beta, the higher the required rate of return from an asset. And the postulated relationship between the beta and required rate of return is linear. This required rate of return can also be seen as a risk-adjusted discount rate for calculating an asset's present value.

From the standpoint of this chapter's analyses, the problem with the CAPM approach is that, for any given level of an asset's relative risk, the risk-adjusted discount rate is constant. However, the argument made here is that, for a given level of asset risk, the risk premium in the discount rate should decline with lengthening payoff period. Another problem with both CAPM and WACC is that neither is amenable to the argument that the risk premium should be zero or less for discounting risky costs uncorrelated with revenue risk.

Conclusions

Discounting risky benefits and costs with risk-adjusted interest rates is at best a crude procedure; ideally we should be maximizing expected utility or applying prospect theory (see Tversky and Kahneman, 1992). However, lacking adequate data for more sophisticated approaches, public and private investors commonly evaluate projects with real, risk-adjusted discount rates often 4 or more percentage points above the risk free rate.

There is no simple answer to the question of discounting future risky values. Unless in some unusually risky situation, such as areas with repeated fire or insect problems, there is support for using risk-adjusted discount rates for long term forestry projects lower than typical industrial rates of return, provided that one is discounting true expected values of revenues and not optimistic projections. Portney and Weyant (1999) also raise a number of intergenerational equity questions about using high short term discount rates for evaluating costs and benefits in the distant future. Industry analysts understandably hesitate to publicly cite discount rates they use for evaluating investments, but some analysts have mentioned to me real after-tax discount rates as low as 4 percent to 8 percent for long term forestry projects. For long payoff periods, typical income tax rates impose a relatively small reduction in rates of return (Klemperer 1996, p. 275). Thus, on a real, before-tax basis, the lower rates in this 4 percent to 8 percent range are below the industrial mean 7 percent estimated by OMB before-tax rate of return indicating that practitioners may indeed be reducing risk premiums in discount rates as payoff periods lengthen.

On the issue of discounting risky costs, a major unresolved question is, by how much should the discount rate be reduced below the risk-free rate when costs are uncorrelated with revenues? The answer depends on factors like the degree of risk aversion, the amount of risk, the number of years that risky costs are discounted, and the level of the risk-free rate. More definitive conclusions await results of further research. Meanwhile, however, the risk-free discount rate, estimated here at a real 3 percent, could serve as the *upper limit* of the discount rate for risky costs uncorrelated with revenues

One of the best ways to deal with risk is to estimate probability histograms for all cash flows and then ask an investor to state certainty-equivalents of these. Certainty-equivalents can be discounted with a risk-free discount rate to arrive at a correct net present value for that investor. But this approach takes more data, expertise, and time than may be available. Using risk-adjusted discount rates is not the best way to account for risk, but since it is one of the easiest and most widely used, it is important to refine the technique. Until we have better approaches that are just as simple, we need not feel that we are misallocating resources if we use lower discount rates for many long term forestry revenues than we use for short term risky gains.

References

Arrow, K. J. and R. C. Lind. 1970. Uncertainty and the evaluation of public investment decisions. *American Economic Review*. 60(3):364-378.

Brigham, E. F. and L. C. Gapenski. 1991. *Financial Management - Theory and Practice (6th edition)*. The Dryden Press. Chicago. 995 pp. plus app.

Brown, S. 1983. A note on environmental risk and the rate of discount. *Journal of Environmental Economics and Management* 10:282-6.

Celec, S. E. and R. H. Pettway. 1979. Some observations on risk-adjusted discount rates: a comment. *Journal of Finance*. 34(4):1061-1063.

Chen, H-Y. 1967. Valuation under uncertainty. *Journal of Financial and Quantitative Analysis*. 2(3):313-326.

Conroy, R. and M. Miles. 1989. Commercial forestland in the pension portfolio: the biological beta. *Financial Analysts Journal*. Sept./Oct.

Dixit, A. 1992. Investment and hysteresis. *Journal of Economic Perspectives*. 6(1):107-132.

Fortson, J. C. 1986. Factors affecting the discount rate for forestry investments. *Forest Products Journal*. 36(6):67-72.

Haight, R. G. 1991. Stochastic log price, land value, and adaptive stand management: numerical results for California white fir. *Forest Science*. 37(5):1224-1238.

Hirshleifer, J. 1964. Efficient allocation of capital in an uncertain world. *American Economic Review*. 54(3):77-85.

Kahneman, D., J. L. Knetsch, and R. H. Thaler. 1990. Experimental tests of the endowment effect and the Coase theorem. *Journal of Political Economy*. 98(6):1325-1348.

Klemperer, W. D., J. F. Cathcart, T. Haering, and R. J. Alig. 1994. Risk and the discount rate in forestry. *Canadian Journal of Forest Research*. 24(2):390-397.

Klemperer, W. D. 1996. *Forest Resource Economics and Finance*. McGraw-Hill. New York. 551 pp.

Lewellen, W. G. 1977. Some observations on risk-adjusted discount rates. *Journal of Finance*. 32(4):1331-1337.

Mills, W. L. and W. L. Hoover. 1982. Investment in forest land: aspects of risk and diversification. *Land Economics*. 58(1):33-51.

Modigliani, F. and M. H. Miller. 1958. The cost of capital, corporate finance and the theory of investment. *American Economic Review*. 48(3):261-297.

Modigliani, F. and M. H. Miller. 1963. Corporate income taxes and the cost of capital: a correction. *American Economic Review*. 53(3):433-443.

Nordhaus, W. D. 1974. The falling share of profits. In: Okun, A. M. and G. L. Perry, Eds., The Brookings Papers on Economic Activity. The Brookings Institution, Wash., D. C. pp. 169-217.

Portney, P. R. and J. P. Weyant, eds. 1999. *Discounting and Intergenerational Equity*. Resources for the

Future Press. Washington. 186 pp.

Prince, R. 1985. A note on environmental risk and the rate of discount: comment. *Journal of Environmental Economics and Management*. 12:179-180

Redmond, C. H. and F. W. Cubbage. 1988. Portfolio risk and returns from timber asset investments. *Land Economics*. 64(4):325-337.

Robichek, A. A. and S. C. Myers. 1966. Conceptual problems in the use of risk-adjusted discount rates. *Journal of Finance*. 21:727-730.

Samuelson, P. A. 1964. Principles of efficiency: Discussion. *American Economic Review*. 54(3):93-96.

Shogren, J. F., S. Y. Shin, D. J. Haynes, and H. B. Kliebenstein. 1994. Resolving differences in willingness to pay and willingness to accept. *American Economic Review*. 84(1):255-270.

Simon, J. L. 1990. Great and almost-great magnitudes in economics. *Journal of Economic Perspectives*. 4(1):149-156.

Stockfisch, J. A. 1969. The interest rate applicable to government investment projects. In: Hinrich, H. H. and G. M. Taylor, eds. *Program Budgeting and Cost Benefit Analysis*. Goodyear Rubber Co. 420 pp.

Tversky, A. and D. Kahneman. 1992. Advances in prospect theory: cumulative representation of uncertainty. *Journal of Risk and Uncertainty*. 5(4):297-323.

U.S. Office of Management and Budget. 1992. *Guidelines and Discount Rates for Benefit-Cost Analysis of Federal Programs*. Circular A-94 (revised October, 1992). Executive Office of the President, Washington, D.C. 24 pp.

Webb, J. F., T. J. Ebner and M. T. Shearer. 1987. Risks associated with timberland investments and the timing of the investment to minimize the risk. Paper presented at the Forest Products Research Society Conference, "A Clear Look at Timberland Investment," April 27-29. Milwaukee, Wisconsin.

Zinkhan, F. C., W. R. Sizemore, and G. H. Mason. 1992. *Timberland Investments - a Portfolio Perspective*. Timber Press. Portland, OR. 208 pp.

ADMINISTRATING RISK – A SOCIAL SCIENCE PERSPECTIVE ON NATURAL HAZARD PREVENTION BASED ON AN AUSTRIAN CASE STUDY

Gerhard Weiss[1]

ABSTRACT

This article discusses how the bodies of public administration deal with natural hazards in theory, and presents the results of an empirical study on policy instruments for natural hazard prevention in Austria. It is argued that different groups have different conceptions about risk. In Austria, two institutions are active in maintaining forests which protect against natural hazards. They apply different routines: the forest authority follows a vague concept of multifunctional forest management. The torrent and avalanche control service, in contrast, focuses more specitically on the reduction of certain hazards. While the first is primarily oriented at the forest owners' interests, the latter concentrates on the interests of the beneficiaries of protective forests. The applied risk policies are strongly determined by the different stakeholders' interests, who participate in the corresponding policy-networks. Within different institutional settings, official experts develop different risk concepts and consequently different solutions.

Keywords: *protective forests, risk policies, public administration, political analysis.*

---

[1] Institute of Forest Sector Policy and Economics, Agricultural University Vienna. – This article is based on a 4-year research study on "Evaluation of policy instruments for natural hazard prevention in Austria" (Weiss, 1998a; 1999). This national project was financed by the Ministry of Agriculture and Forestry and was part of the European COST Action E3 "Forestry in the context of rural development" (Koch/Rasmussen, 1998). The author participates in the EU concerted action MUFOMA "Multifunctional forest management in mountain regions–Evaluation of silviculture and policy means" (http://www.forst.uni-muenchen.de/LST/WALDBAU/mufoma.html)

K. von Gadow (ed.), Risk Analysis in Forest Management, 171–200.

Introduction: The role of forests in natural hazard protection

The relation of forests and natural hazards is discussed in the literature concerning two different aspects: The damage that natural hazards may cause in forests; and the contribution of forests in natural hazard protection for settlements, infrastructures, cultivated land, etc. The first is a management task concerning primarily the forest owner. The latter refers to a relation of the forest (owners) to other parties. These two tasks are basically different problems but still they are interrelated: Natural disturbances in forests may provoke dangerous processes starting from the forest. If a forest is struck by wind-throw and/or a bark beetle attack both timber production and the protection of neighbouring properties may be affected. Foresters around the world have to deal with the protective effects of forests for neighbouring properties or for society as a whole; these are called "protective functions". In some countries the administration of the policy field "natural hazard protection" is assigned to the forestry administration.

In my article I am dealing with the role forests play in protecting mountain areas against natural hazards. As the protective effects of forests concern different parties – the forest owners and the beneficiaries – this is a matter of public policy.[2] For this reason, public administration plays an important role in this field, and I will especially focus on this. The key question of my article is: How do public agencies deal with natural hazards? For finding answers I will use theories from interdisciplinary and sociological risk research as well as from policy and bureaucracy studies, and I will use findings of detailed case studies from Austrian policies concerning mountain forest management and natural hazard protection.

About nature: Are natural hazards different from technical hazards?

When the mass media report about natural hazards they ask "who is responsible?"[3] Usually no case of corruption can be found and nobody has acted illegally. In spite of this, people are injured or even killed and houses and infrastructure are regularly damaged by elementary forces like avalanches, land-slides, torrents or debris flows. Often it is simply concluded that

---

[2] The protective effects of forests are in many cases characterised as external effects and public goods which makes private problem-solving difficult. The basic problem structure is described in Glück/Weiss (1997) and Ottitsch/Weiss (1998), more detailed in Schmiderer/Weiss (1999).

[3] E.g. the weekly magazine *Der Spiegel*, Hamburg, issue 31/97, asks on the cover: "The big flood – natural disaster or work of man?" The cover of the weekly *News* magazine, Vienna, issue 8/99, is titled: "The avalanche. The disaster of Galtür: Who is responsible?"

life in the mountains is dangerous.[4] Even the most sophisticated technology is limited, they say, when facing nature and its unpredictability. Nature can never be controlled 100% and therefore catastrophes by elemental forces will always occur.

Is this true? Are natural hazards different from technological hazards because of the unpredictability of nature? No, there is basically no difference between technological risks and risks caused by elementary forces. Risk always entails a natural process in combination with the utilisation of a natural resource. In the case of atomic engineering, nuclear physics is utilised and it is not possible to control these forces 100%. Accidents occur because the behaviour of water, concrete, heat and radioactivity cannot be predicted precisely. Technological failures always involve human and physical aspects. This is the same with natural hazards: People develop mountain areas, but the natural processes of the landscape cannot be 100% controlled. The specific moment when an avalanche starts and the detailed area it will cover cannot be predicted for each event - but the hazard is usually known. The causes for damage are in both cases neither only human intervention nor only natural processes.[5]

Mountain areas are developed with the awareness of natural hazards, i.e. with a perception of risk. The risk is dealt with by regulations and public agencies. Instruments exist to assess risks, and procedures exist to control development in hazard-prone areas. It is not nature threatening the powerless and unaware population but people utilising nature and endangering themselves, or – to put it differently – they take the risk. It is a lame excuse to state that damage may always occur. It is true that nature cannot be controlled 100%, but it is also true that in principle the risk is – more or less – known and regulated. This leads to the following questions: "Who knows about the risks and who takes decisions? How is risk knowledge managed and how is it considered in decision-making?"

What is risk? Defining and evaluating risk

Decisions, actors and values: the subjective qualities of risk

Risk – this simple word causes ardent public debates and has initiated extensive interdisciplinary studies over the last decades. What is risk? The term itself does not cause much controversy. It is a technical expression like e.g. "duration" or "costs". Describing a

---

[4] E.g. the comment on "Avalanches" in the *Frankfurter Allgemeine Zeitung*, Frankfurt, issue 25. 2. 1999, p. 1.

[5] This lesson is reflected by Jörg Blech in *Die Zeit*, Hamburg, issue 32/97, p. 2, who concludes: "The real reasons for the flood disaster sound banal: there was heavy rain, and people settle too close to the river."

measure as risky is no more exciting or threatening than the attributions "long" or "expensive". Risk is normal: each and every decision contains a certain risk because every decision concerns the uncertain future. The term risk does not evoke anything as long as it is not connected with a specific matter. Mountain climbers or stock-brokers may take high or low risks. Their decisions are only understandable when considering the (subjective) benefits they expect. Discussions about high or low risks are only relevant if it is known what risk and what benefits are concerned. Risk management is usually taken as aiming at the reduction of risk. However, the decision of taking a measure or not is always risky: if the option is taken there is the risk of not reaching the goal; if the option is neglected there is the risk of having missed a chance (Evers/ Novotny, 1987: 32ff, 59ff; Wildavsky, 1989). It is not possible to escape this dilemma that is inherent in every decision. So what is the risk debate all about? In order to answer this question, I want to explain the following basic features which are essential to understand "risk":

Risk is connected with decisions: Risk has to be distinguished from danger (Evers/Novotny, 1987: 32-46; Luhmann, 1993). The concept of risk goes back to medieval trade and is formalised in decision theory. Only the possibility to choose between options sets up a risk situation; if I accept a risk I also take a chance. The choice constitutes the probability of profit or loss (Müller, 1993: 3813f). If I cannot choose I am endangered.

Risk concerns actors: The fact that risk is connected with a decision points to the fact that there is an actor who decides. In formal decision theory this actor is abstract; it presupposes a "rational decision-maker". In real life decisions, however, the "subjectivity" of all actors is crucial (Evers/Novotny, 1987: 42). There is more than one rationality – it is crucial who decides.

Risk includes values: According to decision theory, decisions on risks weigh present and potential future benefits and costs and their probability. In real life decisions, benefits and costs are weighed by actors with interests and values. Decision models are constructions that can either meet the real situation of the deciding person or not. Risk concepts are not neutral but include normative assumptions about the decision situation. Economists state that it is impossible to define an "objective" rationality that can generally be applied (Fritsch/Wein/Ewers, 1993: 148).

Decisions about the safety of technical projects are usually based on risk analyses or risk assessments. Risk management tries to find a rational approach to safety decisions by structuring the problem and making the decision transparent.

These rationalist risk models, which are developed by technicians, are often not accepted by the public. Prominent examples for conflicts between experts and laypersons are the discussions about nuclear reactors or genetic engineering. The concepts of risk analysis and risk management were applied in the field of natural hazards as well (Kienholz, 1993; Hollenstein, 1995). These authors recognise the problem of deciding "What is acceptable risk?" They propose to include risk preferences in the assessment process (compare Starr, 1969), but these attempts under the psychometric paradigm were not successful either. In varying situations people accept differing risk levels. The risk preference of one person in a certain situation cannot be transferred to another person in the same situation, nor can conclusions be drawn from the risk preference of one person in a certain situation to other situations. The same is true for different social or cultural groups (Johnson/Covello, 1987; Jungerman/Slovic, 1993).

Technicians may recognise that the question "How safe is safe enough?" cannot be answered by expertise alone, but has to be answered by politics or society. However, it is common that politicians fail to give clear directions about an acceptable level of risk. Kienholz and Hollenstein mention the problem of "political" or "subjective" factors in risk evaluation, but up to now conceptual approaches to cope with these problems have been lacking in natural hazard regulation. Risk managers try to deal with these questions within their technical concepts, e.g. by including risk preferences into the decision model. The questions concerned, however, go beyond the technical concepts. They are questions of political procedures. The bureaucratic decision and the use of risk models is only one possible procedure. These existing procedures put value decisions into the hand of technicians who normally are not aware of the political implications. In many risk debates this technocratic procedural approach is questioned (Nelkin/Pollack, 1980: 243-246).

The traditional bureaucratic approach applied by public agencies has deficiencies in terms of democratic principles because the interests of different groups are not considered evenly in the implementation process. In spite of their formal rationality the concepts applied are often biased towards powerful interests (Mazur, 1985). In a study of the practical application of risk assessment the questions of power distribution and the role of political actors have to be included (Compare, 1995). In practice, risk models may not be adequate in certain situations at all or they may be manipulated or instrumentalised by certain groups. These problems are also found in natural hazard protection (Weiss, 1999). Alternative procedures – e.g. public participation – have not yet been studied thoroughly.[6]

Actors, values and interests: conceptions and politics of risk

Risk is technically defined as "the probability and magnitude of an adverse effect" (Lowrance, 1980: 6). The objectivist paradigm in risk research believes that risks may be assessed objectively. Technicians work with instruments like risk analysis and they usually assume that "rational actors" will follow their concept. In contrast to this, risk researchers who approach the problem with a constructivist view state that risk is constructed in the mind of people – and is thus constructed differently by different people. From their point of view, an objective definition and assessment of risks is not possible (Douglas/Wildavsky, 1982).

A possible answer to the conflicting objectivist and constructivist theories is that both views are complementary (Rosa, 1997). It can be assumed that real hazards exist; in the perception of people, however, they are always constructed (Johnson/Covello, 1987). In such a way also the technical, objectivist concept of risk is one cultural form of risk perception (Krohn/Krücken, 1993: 13). Not only the evaluation of risks is based on "cultural specifics" (Kienholz, 1993: 18) but also its conception.[7] The attempt to understand risk debates leads to the necessity of a broader view on the subject. To understand the risk debates the subject has to be framed in a social context. It is useful to take a look at risk concepts from a meta-analytical level: to look how risk concepts are formed. We can ask: "Who defines risk in what way?" The probabilistic risk model is just one possible conception of risk among others. It is connected with our modern society, originally being generated in the course of the individualisation process that started at the beginning of modern times (Elias, 1976).

What makes risk politically relevant is that different people may be affected by risk decisions. It is the actor assessing the risk who is crucial! The subjective character of risk assessment becomes clear when considering that risk decisions do not only include the "risk

---

[6] Recently a study about the options of public participation in torrents and avalanche control projects in Austria has been carried out (Ottitsch/Pregernig/Rappold, 1998). This topic is considered in ongoing studies in Europe ("Public participation in mountainous ecosystem management", EFI; EU concerted action "Multifunctional forest management in mountain regions – Evaluation of silviculture and policy means", MUFOMA).

[7] In the literature different risk conceptions as well as different concepts of risk are debated; Shrader-Frechette (1995, 1997) distinguishes positivist and relativist conceptions of risk, whereas the framework developed by Thompson and Dean (1996) proposes a dichotomy of probabilist and contextualist risk concepts. For the purpose of this article it is sufficient to recognise that "risk" does not mean the same for different people.

factor" but also potential costs and benefits. It is plausible that different actors may value things differently because of personal preferences. It is even more obvious that people judge risks differently depending on how they are affected by the decision: Do they benefit from the measures? Do they have to bear the costs? In the cases of nuclear power plants and genetic engineering in the first place the companies profit but the public bears the risk or the potential costs. A similar situation is found in the case of developing hazard-prone areas in mountainous areas: The property owner makes profit from hotels or shops but if torrents or avalanches destroy the building other people are injured or killed and parts of the costs are covered by the public. The dissent between laypersons and experts is different in the case of the power plant and the case of natural hazards: In the first situation, the official experts typically share the view of the power agency and laypersons oppose. In the latter the experts represent public interest in restricting the development of hazard-prone areas but laypersons are interested in development as property owners. The main reason for the diffrence is the relative power of the interest groups.

The dissent of laypersons and experts is frequently explained by limited information and insight of the laypersons. Experts tend to believe that their view on risks is objective and the public perceives risks differently – i.e. in a distorted way (Starr, 1980: 4). Slovic/Fischhoff/ Lichtenstein (1980: 212), however, show that experts are not immune to biases in risk percep-ion. Laypersons' perceptions are often considered "irrational" but their behaviour can also be called "rational in a different sense". On the contrary, one could even say that laypersons behave more rationally than experts in certain cases. In the example of nuclear power plants Perrow (1984) shows that experts failed to recognise the specific susceptibility of the technology to accidents but laypersons did. In this case, relying on limited calculations and working with probabilities by experts can be considered irrational; the sceptic attitudes of laypersons with regard to unpredictable consequences of accidents seems rational. The dissent between laypersons and experts can partly be explained by clarifying what is understood by risk. The risk model of experts is limited to probabilities of a certain damage. Studies show that layper-sons often assess probabilities of different hazards quite accurately (Lichtenstein *et al.*, 1978; Renn, 1984). Different results are obtained if we do not establish death rates but simply ask "how risky" is a certain activity. People call sources of risk more or less risky quite independent from their own assessment of death rates or damage (Renn, 1984: 152). That means that in their risk assessment qualitative aspects are included additionally to quantitative information. It is important to recognise that risks are not only evaluated differently but also constructed differently. There are not just different preferences but different models behind the decision. In the case of natural hazards it will make a difference if the property owners who are interested in development are asked about their opinion or inhabitants who strive for maximum safety. The land owners would criticise other aspects of official risk assessments than other persons.

Finally it has to be stated that the public risk debates were about more than just risk: they were about technology, about democracy, about equity, about the environment (e.g. Nelkin/Pollack, 1980: 243; Beck, 1988: 20). In this respect the risk debate can also be interpreted as a ritualised confrontation between social or political groups (Conrad, 1989: 183-190). The attractive topic of risk may hide other topics like the criticism of technologies, but it also hides other latent conflicts concerning the application of these technologies (Conrad, 1989: 190-194). If problems are concerned where the concepts of risk and insurance are no longer adequate the symbolic risk debate distracts from the real problem. In natural hazard prevention such cases may be found concerning "close-to-nature" prevention concepts. Conservationists' criticism of e.g. a check-dam may go beyond the impact of this single construction but may question the whole prevention concept.

At the beginning of this chapter I stated that it is the context which makes discussing risks interesting. It makes a difference what kind of risk, what kind of activity is concerned, and it makes a difference who assesses the risk. Three important conclusions may be drawn from this chapter for analysing risk:

- Risk assessments always involve values and interests of groups. Therefore, the question of power may not be excluded when studying risk debates. Risk decisions include (potential) costs and benefits and therefore it is crucial which groups will have to bear the costs and who profits.
- Different social or cultural groups not only evaluate risks differently but also have different perceptions and conceptions of risks. For instance, dissent between laypersons and experts may not only occur because of varying assessments of the probability or a varying evaluation of damage but also because of varying basic assumptions about and different aspects considered in the assessment.

- Public debates about risk may explicitly or implicitly concern additionally other topics connected with the activity in question, e.g. the use of technologies, impacts on the ecology, democratic processes, etc.

Forests, natural hazards and risk assessment

Who assesses the risk in the field of forests and natural hazards? Important groups of actors are: forest owners, direct beneficiaries, the general public, and public administration. The owners of properties from which dangerous processes may start (e.g. *forest owners*) have direct influence on the risk situation by their regular management decisions. The question here is: "In which way, if ever, do they consider the risk situation in their decisions and how can this consideration be retained?" Answers can be found in their value system and their interests. Appealing to their solidarity may make them consider the safety of people living below their forest. Forest owners may feel responsible for possible damage; however, they try to reduce costs for protective measures because they do not benefit from the safety (unless they own the properties below themselves). Their liability in case of a damage possibly provoked by their management measures is an interest-related factor determining management decisions of forest owners. The owners of properties potentially endangered by natural hazards, or benefiting from protective effects of forests (*direct beneficiaries*), are directly interested in reducing the hazard. They are interested in the highest possible safety for the lowest possible costs. The *general public* is interested in many positive effects of the forest but latent interests of big groups are difficult to organise (Olson, 1965). The *public administration* tries to ensure the public interest, which is sustainable forest management including the maintenance of the protective effects of the forest. Public bodies formulate and implement regulations for forest management and natural hazard protection. One important part of this regulation is the definition of property rights in protective forests: "Is the forest owner obliged to maintain the protective effects of the forest or not?" And another question is: "How is the regulation implemented?" I want to discuss this question in the following section.[8]

Property rights and the public administration form the institutional background for the policy of natural hazard protection. The basic determinants for the behaviour of the forest

[8] The question of property rights is not elaborated further in this article. Detailed results on the behaviour of owners of protective forests, beneficiaries and public agencies can be found in Schmiderer/Weiss (1999).

owners was shortly explained above. The following chapter focuses especially on the behaviour of the bureaucracy.

To administer risk: The behaviour of bureaucracy in risk regulation

Public agencies as political actors

Public agencies play a major role in risk policy because they are main actors in both the formulation and the implementation of norms, concepts and procedures for risk regulation. Policy formulation is usually not done by parliament independently from executive bodies. It is common that the administration elaborates proposals for sector laws. By doing so, not only their knowledge, but also their value system and interests are incorporated in the regulation (Peters, 1984). Furthermore, they are the main implementing actors. Laws are formulated on a general level. Very often policy goals are formulated in the law only vaguely or even contradictory. Conflicts between interest groups that have not been solved in the formulation process lead to vague formulations, which hide underlying conflicts. They then have to be solved in the implementation process. If strong interest groups still oppose regulation at this stage, laws cannot be implemented. Implementation deficits are the result (non-decision problems, Wolf, 1996: 145, 151).

In the implementation process, general norms are applied by the administration which thus strongly influences the interpretation of the laws. However, they are not independent in this process: actually, there is an implementation network in which other actors are involved as well. Through this network, values and interests of parties also affected by the regulations become effective in implementation. Each group pursues their interests, and the result of the implementation process depends on the power of the actors involved in the network. As a result, risk concepts and risk evaluation of experts and official agencies are biased. Risks are conceptualised within the value system of the profession or the policy community, respectively (Dietz/Rycroft, 1987: 77). Technicians, for instance, tend to describe risks mathematically. Depending on influential interest groups different criteria may be included in the evaluation: economic value of damage, magnitude of possible catastrophes, kind of damage expected, number of deaths expected, impairments for quality of life, short-term and/or long-term effects, ecological damage, etc.

Informal interests, professional values, and safe decisions

To understand public bureaucracies one has to be aware that not only the formal goals (contents of laws, general principles of bureaucracy like legality, economy, expediency) but also the informal goals determine their behaviour (e.g. Downs, 1967; Mayntz, 1985; Scharpf, 1997). Such informal goals which are significant for every organisation, are for instance: maximising their budget or safeguard their autonomy, competence field and legitimacy. Public agencies are often characterised by a clientelistic behaviour: In order to minimise conflicts they try to regulate a sector rather in correspondence with than against the interests and values of their main clientele (selective clientelism). Measures are negotiated between the administration and the addressee (bargaining process). The result of this negotiation process depends on the power of the groups concerned: If the private actors concerned by a regulation are powerful (e.g. if they have a significant economic status, a high number of employees, etc.) the result may be that the public agency serves the interests of the private group rather than controlling them (capturing, Wolf 1986; "state failure", Jänicke, 1986). Clientelistic behaviour like in industry is also found in the forestry sector (Glück, 1976; Krott, 1990; Pregernig/Weiss, 1998). In environmental policy within the industrial sector clientelism leads to "end-of-pipe" solutions (Wolf 1986: 158). A change of structures and processes is not achieved because of the resistance of established actors which benefit from the existing situation. Typical environmental problems are externalisations of costs; an internalisation is defended by the system (Jänicke, 1986). In natural hazards policy this means that, for instance, constructions to defend damage by torrents and avalanches are preferred to afforestations (Weiss, 1999).

The realisation of measures depends not only on the "objective" problem pressure but also on the problem solving capacity of the administration. Prittwitz (1990) describes the 'catastrophe paradox' as a phenomenon that if the problem solving capacity is restricted, small problems are solved whereas bigger problemsremain unsolved. In the work of a bureaucracy even a reversal of the rational principle of choosing means according to certain goals might occur: The administration may have certain instruments at hand and, with the informal goal of keeping its territory and legitimacy, they seek goals to legitimise their work (Luhmann, 1966; Downs, 1967; Edelmann, 1976). By applying their routines technicians would look for technical solutions, foresters contribute by silvicultural measures. Eventually measures are taken in a way and at locations that can be realised without great troubles (aversion of conflicts). In their problem solving efforts they apply their professional values and routines and change policies only under high pressure (Sabatier, 1986, 1993).

For public agencies the problem of risky decisions is not a new situation: all decisions include risk because decisions are always taken under uncertainty. For the administration, however, decisions have to be certain. Procedures serve to transform the uncertainty that lies in the problem into certainty of the decision (Luhmann, 1993: 144f). In the case of environmental risk, setting standards of maximum emissions, measuring emissions, and mathematical procedures rationalise the highly uncertain and value-laden decision on "what is a health hazard" or "what is acceptable risk" (Wolf, 1986). Formal procedures make the decision appear value-free. Further more in the assessment of natural hazards there are procedures that are ostensibly "objective". Yet, the underlying normative assumptions of what risk is referred to as acceptable is hidden behind expertise.

Two agencies - two concepts: Case studies on risk regulation in Austria

The forestry administration in Austria deals with classical forestry tasks as well as the protection against torrents and avalanches. Beside the forest authority an agency was established under the forestry department, other ministry of agriculture, especially for the task of natural hazard protection (torrent and avalanche control service, TACS).[9] Partly, the same policy means are applied by both institutions. This allows a comparative analysis of two institutions active in the same field. By this comparison the importance of the institutional design in regulating a policy field can be illustrated. In the following I will present the application of exemplary policy instruments by these two institutions and draw conclusions with regard to the determinants of their behaviour in risk regulation. All findings are the result of a research study which I conducted in the years 1994-1998.[10]

---

[9] Although there is a federal law as against provincial laws on forestry, there are forest authorities for each province. To simplify matters, I will speak of the forest authority in singular. By this I mean the provincial authorities and the ministry of agriculture, forestry department. The TACS on the contrary, is directly subordinated to the ministry.

[10] The examples used originate from the research project "Evaluation of policy instruments for natural hazards prevention in Austria" (Weiss, 1998a). Aspects on the administration of risk are partly included in the dissertation which is based on this material (Weiss, 1999). In this project, data was gathered at federal and provincial governments (interviews, statistics, documents) as well as at the local level. In 20 districts the local officers of the forest authority and the TACS were interviewed (representing about ¼ to ¾ of the districts in the mountainous parts of Austria) and 50 projects were chosen for detailed analysis (interviews, document analysis and visits at site). The results can be regarded representative for the administration of torrents and avalanches in Austria.

The forest authority

Protected vs. protective forests: the politics of definitions

The formal institutional goals of the forest authority are the public interests in the forest, which are in Austria legally defined as the effects of the forests on wood production, natural hazards protection, protection of environmental goods, and recreation. The fact that 80% of all forests in Austria are privately owned has a strong effect on forest policy, because of strong forest owners' interest groups. For instance, most subsidies are granted for a better utilisation of the forest by the owner (so-called "production function"). Over the last three decades, however, activities of the forest authority in "protection forests" has increased gradually (so-called "protection function").

Traditionally, Austria's mountain forests are regarded as "protection forests" (Schutzwald) implying a special protected status for these areas. The term "protection forest" comprises two aspects: (a) the protection of the forest against over-use or improper management, and (b) the importance of the forest in protecting the soil against erosion. Going back to the Austrian Empire's Forest Act of 1852, the Austrian Forest Act of 1975 defines "protection forests" as site-protective forests. They do not necessarily have direct protective effects for settlements or infrastructures against natural hazards. Protection forest/Schutzwald is an ambiguous term: While the official definition refers to site protection, in colloquial use (also by foresters) it is rather connected with the protection against natural hazards. This ambiguity is used politically by the forest administration: Statistics are based on the legal definition which results in large areas of protection forests (most parts of mountain forests); in public relations, however, it is argued that these forests are necessary for protecting against natural hazards. In fact, only few forests directly protect against rock-fall or avalanches. In their mind Austrian foresters typically do not distinguish between forests at risk and forests as part of a risk reduction scheme. This undifferentiated connection of protection forests and protective effects of the forest in their colloquial use influences the practical work of the forest agencies: They are hardly aware of different protective effects of the forest and have no corresponding management concepts at their disposal.

"Multifunctional" protection forest restoration

The traditional protection forest policy was a "do-not-touch" policy, which is understandable when considering the "classical" foresters' fears of over-exploitation. In the last decades a new problem definition and a new paradigm in protection forest management evolved: As timber prices fall, revenues from wood production decrease. At many sites, especially in mountain areas, wood production is no longer profitable. Instead of over-use, "under-use" or "lacking management" of mountain forests is considered a problem today. This view serves the interests of the authority which – facing the decreasing economic significance of the forestry sector – searches for new tasks to maintain its own legitimacy. The forest authority initiated new subsidy programmes for "protection forest restoration". Although these programmes were officially argued with the protective effects of forests against natural hazards the measures focused on protection forests according to the Forest Act. Only a small part of these forests actually provide protection against natural hazards as well. When granting subsidies under these programmes the forest authority does not primarily consider direct protective effects against rock-fall, avalanches, etc. Practically all forest management measures carried out by forest owners are subsidised, and measures specifically oriented at an enhancement of protective effects are hardly considered. The forest authority assumes that all "forest functions" are enhanced in an indirect way by regular forest management. Furthermore they hope that protective effects are safeguarded by supporting economically oriented forest management, which is carried out by the owner. Multifunctional forest management is not operationalised according to the different "functions".[11] Thinking in the tradition of the "wake theory" it is believed that orderly forest management automatically fulfils all forest functions.[12]

Perceptions of risk

In subsidised projects for protection forest restoration in Austria, wind-throw areas are usually treated traditionally, which includes the harvesting of the timber followed by afforestation. This treatment produces bare ground like clear-cuts, thus increasing the threat of avalanches and erosion. In many cases leaving the thrown timber at the site would have been cheaper and more favourable in terms of protective effects. Why is the clearing of the wind-throw sites promoted by the forest officials? Clear-cuts are traditionally regarded as normal treatments, and therefore accepted by forest officials as well, although they are unfavourable in terms of natural hazard protection. Clear-cuts are assessed positively because they are intentional actions; wind-throw

[11] The forest policy doctrine of forest functions goes back to Dieterich (1953). In practice, the term "forest function" is rather to be understood as an ideological than a scientific expression (Heeg, 1973; Glück/Pleschberger, 1982; Krott, 1985).

[12] The ideology of the so-called "wake theory" seemingly harmonises competing interests in the forest (Glück/Pleschberger 1982: 652).

on the other hand is regarded as damaging - as nature disturbing planned forest management. The language chosen by forest officials shows how management measures and natural 'catastrophes' are perceived: wind-throw areas are often called areas 'cleared' by storms, although the storm alone did not actually clear the ground - this is eventually done by the forest workers who remove the timber from the site. It was not the natural incident but human intervention that produced bare ground. Foresters traditionally fear risks if wind-throw sites or even healthy forests are left untreated.

Forest officials and silviculture scientists in mountainous countries in Europe usually appeal for more active management of mountain forests (e.g. Mayer, 1967a, b, 1976; BMLF, 1993). They argue that the natural development of forest stands includes unstable 'stages' such as the 'decomposition stage' - when old trees die and regeneration starts - susceptible to natural threats. Another view - that natural mountain forests are particularly stable ecosystems (e.g. Meister, 1985) - is shared by some foresters and environmentalists. They consider forest management destructive to mountain forests, for example producing unnatural and unstable stands susceptible to wind-throw, insect attacks and other threats. Both views have firm arguments; both are based on scientific knowledge and practical examples may be found for both positions; the generalisations, however, are guided by ideological beliefs that can be interpreted as different views of nature. Nature may be seen as harmony or as chaos (Sieferle, 1997: 19-21). In the first view human interventions cause risks, in the latter natural forest dynamics do (Weiss, in press). In their "silvacentrist" view (Glück/Pleschberger, 1982) foresters do not differentiate between risks for the forest and from the forest. For them risks for the forest are at the same time risks for society.

Recent research indicates that many cases need less intensive interventions in mountain forests than foresters tend to believe – even forests damaged by wind-throw and bark beetles (Frey *et al.*, 1995). In certain cases, however, treatment seems to be necessary to avoid unacceptable levels of risk, e.g. on very steep slopes directly above settlements or infrastructures. The actual policy applied is highly determined by the stake-holders involved in the policy-making process: A low-intervention strategy where timber has been left on wind-throw sites was practically applied in certain cases in the Austrian province of Vorarlberg and in the neighbouring countries Switzerland and Bavaria. In these countries the share of municipal forest property is high which gives other interests than timber production more weight in management decisions. Especially in Switzerland nature protection groups are explicitly included in forest policy-making (SBN, 1994).

The torrent and avalanche control service (TACS)

The "forest-technical system"

The institution of the forest-technical service for torrent and avalanche control dates back to the late 19th century when the Act for Torrent Control came into force after severe floods that occurred in large parts of the Austrian Empire in 1882. Since that time the service is assigned to the forestry jurisdiction and implements what is called the forest-technical system of torrent and avalanche control. This concept aims at combining biological and construction measures for preventing natural hazards. Technical means should not only be applied to prevent damage; the land should be used in a way that hazards are avoided as well (precautionary principle). For instance, alpine pastures should not be overgrazed so that no erosion or land-slides occur; alpine meadows should be afforested if this can prevent avalanches; in forest management, clear-cuts that might cause avalanches, rock-fall, or land-slides should be avoided. In recent time another instrument, following a precautionary approach as well, was developed by the TACS: hazard zones mapping. Land-use should not only be controlled in the catchment areas of torrents and avalanches but also in the hazard-prone areas themselves. While primarily agriculture and forestry are concerned in the catchments, the land threatened by hazards are often settlement areas. In prospering times many new buildings were constructed in such areas, thus causing additional needs for prevention measures against hazards. Hazard zones planning is now striving to prevent the development of such areas.

Conflicts in land-use regulation

The TACS carries out biological and construction measures and elaborates hazard zones plans. A large amount of money is invested in prevention measures every year (over 1 thousand million ATS). On average 80% of the project costs are subsidised. Beneficiaries, e.g. municipalities, have to apply for projects; they are the clientele of the TACS. Although the forest-technical system comprises theoretically both land-use regulations and construction measures, actually mainly technical measures are carried out.

The preference for technical constructions is not strictly dependent on weighing system performance or by considering their return on investment. One important reason why constructions are preferred over land-use regulations is that the restriction of land-use causes more conflicts with local users (e.g. afforestation of alpine pastures). Furthermore, technical constructions like check-dams have immediate effects which have a higher political value. Politicians at municipal level prefer immediate effects and are not only interested in physical impacts (less future damage) but also economic (increased property values) and symbolic benefits (proof that they "care" about the safety of their municipality). The symbolic value of constructions is higher because their protective effects are more obvious. Although afforestations would provide a long-term enhancement of the communal safety, they are not promoted adequately. Why – if the TACS considers land-use regulations a central part of watershed amelioration – are these measures not safeguarded more strongly?

Table 1 shows the typical policy network in planning an avalanche prevention project. The acceptance of constructions and biological measures (afforestation) in such a project is indicated with regard to the type of political actors usually involved. Actors participate in the policy network formally or informally. As a rule, not only the village/town (applicant) has high interests in realising technical constructions but also the county authority which informally represents influential regional interests. The constructions themselves are opposed only by nature protection concerns. On the other hand, there are rather powerful actors opposing the afforestation of alpine pastures in such a project (farmers, hunters, agrarian officials, tourist managers) while only a few actors support this land-use regulation (representatives of the ministry, the local forest office and the nature protection office). According to the principles of their "forest-technical system" the local TACS officer usually proposes afforestations wherever ecologically feasible but only few of them pursue this goal if they face strong opposition from local actors.

	ACCEPTANCE **)	
ACTORS *)	Technical constructions only	Afforestations in addition to constructions
Torrent and Avalanche Control Service local office	+	?
Applicant (e.g. commune, transportation agency)	+	0
Land owner (farmer)	0	–
Hunting rights holder	0	–
Tourism	+	–
Regional authority (county)	+	0
Agrarian official expert	0	–
Ministry of Agriculture (Dep. TACS)	+	+
Forestry official expert	0	+
Nature and landscape protection official expert	–	+

*) *Interest groups are ranked according to the significance and power in the local implementation process.*

**) *The actors' interests in implementing the stated measures are valued as: support (+), neutral (0), resistance (-); the question mark (?) means that the position of the actor is „not homogeneous".*

Table 1. Typical attitudes of political actors towards avalanche control projects with and without afforestation (source: WEISS, 1998b: 123).

As the study underlying this article indicates, the result of this distribution of interests is that "technical" projects are usually preferred over afforestation projects, only minor afforestation areas are included in projects, and the success of afforestation is often poor because they are not properly protected against damage by browsing etc. Other interest groups or the public that would possibly criticise unsustainable land-use and thus could strengthen the position of the TACS are not included in the decision-making process.

Bias in risk-benefit analysis

As a prerequisite for granting subsidies for natural hazard prevention projects in Austria, cost-benefit analyses (CBA) have to show a positive result. This tool goes back to an initiative of the ministry of finance. The guidelines were elaborated by scientists under the head of the ministry of agriculture. A cost-benefit analysis is a management tool of the central organisation – it is not in the very own interest of the street-level bureaucrats because it restricts their competence of decision. The policy evaluation shows that the TACS does not accept this instrument as a decision tool at all. Decisions about what projects will be realised are not based on a cost-benefit analysis; this analysis tends to be elaborated when the decision has already been taken unofficially by the TACS regional officer and the measures are already planned in detail. The CBA is just used for legitimising the functionality and efficiency of the project proposed. Decisions are mainly taken in the regional offices of the TACS that choose, project and realise the projects. Effective factors for the practical decision besides of the estimated hazard are (a) the interest and the financial capacity of the applying community, (b) public attention, (c) informal interests of the agency, and (d) the regional economic development or political priorities on provincial level. The impacts of the formal CBA in torrent and avalanche control can be seen in (a) a certain influence on the thinking of the decision makers in terms of efficiency when projecting measures and (b) in sorting out the very least efficient projects.

The shortfalls of the CBA as applied in practice are manifold. There is a wide range in assessing the benefits of such projects. For most projects supported by the agency the CBA can be calculated in a way that the result supports its realisation. The formal need is only to prove the single project "positive"; projects are not ranked nation-wide. The CBA as applied within the TACS reflects neither the economic goals of the ministry of finance nor the official principles of the TACS, either in formulation or in implementation.

Although being claimed an "objective" tool the CBA falls short of this pretension with regard to both theoretical and pragmatic aspects. The CBA as it is applied is not scientifically sound with respect to economic theory because different value categories are simply aggregated.[13]

Neither does it pragmatically reproduce the formal goals of the TACS. The positions considered within the CBA framework do not adequately reflect the official principles of the TACS. Principles formulated by the agency state that (a) the protection of existing facilities should be prioritised over developing new ones, (b) the protection of lives is more important than the protection of material values, and (c) developments should not be supported that

---

[13] Guidelines *("Kosten-Nutzen-Untersuchung")*, BMLF (1978).

result in future needs for increasing prevention measures.[14] There are positions included in the CBA that support projects which are not favourable according to these principles. Land that might be developed due to the project in question is evaluated with its expected future value (estimated rise of the property price). This value is usually several times higher than the estimated damage of existing houses (the damage is devaluated and reduced according to the probability of a damage). Furthermore, the CBA itself does not consider human lives. Representatives of the TACS on the one hand call this a big drawback of the analysis, on the other hand they believe that human lives should not be expressed in monetary terms. They use this situation to argue that the threat of lives is to be included in the decision additionally to the CBA. The hazard zones plans indicating the number of houses possibly affected by natural disasters serve as a database for the consideration of possible deaths. However, the consideration of potential fatalities does not take place formally. As a result, there is a wide spectrum of possible decisions. The agency does not want their discretion to be restricted by a standardised procedure.

The political system undermines the CBA in another way: by using regional economic arguments, even projects might be realised that do not produce a positive cost-benefit relation. The project is then called a contribution to the economic development of disadvantaged regions. Most projects, however, are realised in prospering regions, e.g. regions with tourism. As a result the natural hazards protection policy does not restrict but support economic development in mountain regions. Growing economic activities in these regions mean growing threats and increasing needs for protective measures.

Operative deficits of hazard zones planning

The main policy means of the TACS – prevention measures and hazard zones mapping – are not implemented independently of each other. There are several relations between these instruments. Formally (according to the Forest Act), there are two purposes of the plans: (a) to prevent the development of hazard-prone areas for housing etc., and (b) to provide comprehensive data on the hazardous processes as a basis for systematic planning of prevention concepts. Informally, hazard zones plans also examine the "objective" necessity of prevention measures as a first step. The TACS uses the plans to argue for the elaboration of prevention projects in villages/towns if buildings are situated in "red zones" (high hazard

[14] E.g. guidelines *"Hinderungsgründe"*, BMLF (1980).

zones). In a second step, the plans serve for evaluating the achievements of the service for the safety of the community. The municipalities, who have to cover a share of the project costs, want to "see results". A higher standard of safety is visible in technical constructions and furthermore in changed hazard zones: after the completion of a project, "red zones" are reduced in the maps. What politically counts is not the real reduction of hazards, but the reduction of hazard zones on the map. With the elimination of "red zones" these areas are free for economic development. The value of these properties may increase dramatically, especially if situated in tourist regions.

The problem with this procedure is that risks are partly ignored. The hazard zones maps provide only restricted information; the hazard assessment refers to a "standard or reference event". This reference event is a disaster of a magnitude that is statistically expected every 100 or 150 years. The remaining risk of events with higher impact, the uncertainties of the assessment, as well as the uncertainties of the construction measures are not considered in the assessment and are not shown in the map. Also constructions are laid out for preventing the "standard event" only. If new houses are erected in formerly "red zones", now protected by technical constructions, they are still threatened by remaining risks and uncertainties. These remaining risks are ignored (Jäggi, 1996); the prevention concepts provide "false" or "relative" safety (Weinmeister, 1988). There are examples in Austria where newly built houses were damaged by events that were more dramatic than "expected". In other cases the prevention measures failed (Aulitzky, 1968, 1970, 1996). 100 years ago the reference event for prevention constructions was an event with a 30 years probability of recurrence; 50 years ago this standard was 50 years. What safety will be expected in the future? In history, villages and towns were built in the protection of constructions, the protective effects of which are insufficient according to today's safety standards. Today's measures probably do not meet future expectations. The development of areas protected by technical constructions inevitably results in future damage. The protection gained by the constructions is given away if new houses are erected. Such a policy enlarges the potential development area but does not reduce hazards on the whole.

A joint restoration programme for protective forests

Initiation and formulation

In the late 1980's a joint programme of the forest authority and the TACS was launched to support mountain forest restoration. The ministry – the initiator of this subsidy programme – observed that the TACS prioritised technical constructions over measures aiming at maintaining the protective effects of forests. The forest authority, in contrast to this, was interested in carrying out such forest management measures but the budget of the protection forest restoration programme existing at that time was insufficient. With the new programme money from the Disaster Relief Fund administered by the TACS was made available for the forest authority. For formal reasons restoration projects had to be borne jointly by both institutions. In the following a dispute between the two institutions arose about the objectives of the programme. Officially, the programme was called "restoration of damaged forests in catchments of torrents and avalanches" because as a part of the disaster relief programme it was restricted to natural hazards. Practically, the measures were called "protection forest restoration" by the forest authority because they wanted to implement their restoration concept in "protection forests" according to the Forest Act (see above). In the understanding of the TACS, restoration projects should have the character of watershed management measures oriented at enhancing the protective effects of the forests against natural hazards ("protective forests"). Both agencies applied their routine concepts: While the forest authority wanted to be active in all "protection forests", the TACS considered only "protective forests" as targets of the programme.

The conflict between "protection forest restoration" and "watershed management" was in fact a conflict over competencies and resources of these two institutions. The forest authority insisted on the term "protection forests" according to the Forest Act and the TACS adhered to their definition of "protective forests" in terms of natural hazards. Defining the programme goal with reference to "protection forests" would have granted the forest authority a larger field of activity. Focusing on "protective functions" would have strengthened the TACS. Eventually, a compromise was reached by not generally excluding measures on a larger scale but giving priority to forests with direct protective effects.

Implementation

The obligation to joint projects of the two agencies also caused difficulties because of unclear responsibilities. This problem was solved by defining a main responsible agency for each project. Practically, projects carried out by the forest authority or the TACS differ (see Table 2): The forest authority includes larger areas in the projects and encourages the forest owners to carry out the measures. Measures are planned in such a way that they also meet the interests of the forest owners. The TACS, in contrast to this, focuses on areas directly protecting settlements or traffic lines, and carries out the work on its own. Measures of the TACS are designed for immediate and maximum protective effects.

The main clienteles' interests influence the decisions of the public agencies. This results in a timber production orientation of the forest authority additionally to the protection goal (preferences of the forest owners) and in a short-term orientation of the TACS (preferences of the beneficiaries). While the management concept of the forest authority includes forest management measures also oriented at timber production, the TACS focuses more on the specific protective effects desired.

Two risk concepts

The two institutions operate with different risk and management concepts (Table 2): The *forest authority* aims at a multifunctional forest. They assume that "close-to-nature management" will result in a "stable" forest, i.e. the best provider of all "functions" of the forest. Forest owners are supposed to play an important role in keeping a stable forest. The forest authority's subsidy approach is oriented at the forest owner. Subsidies should be incentives for them to consider public interests in their management decisions. Measures subsidised by the forest authority do not derive from specific protective effects of the forest because they require that mountain forests generally provide a long line of important protective effects for society. As they consider indirect effects just as important as direct effects, in their view the specific protective effects do not have to be proven in each situation. They have a holistic but vague approach. In the view of the forest authority the risk of forests losing their potential to carry out this "function" may be a result of natural disturbances. They look at risks for the forest. If the forest is at risk, its "functions" are endangered as well. Furthermore, the forest they generally have in mind is rather a "timber-forest" than a protective forest because this is the traditional image of the ideal forest.

The *TACS*, in contrast to this, applies a monofunctional management concept. They start with certain protective effects forests have for settlements or infrastructures. Measures derive from specific protection goals related to certain threats. In their view they subsidise measures in the interest of the beneficiaries of the protective effects. Their risk concept is directly connected with the security of the properties below the forest. Their protection concept is derived from certain natural processes, which potentially threaten human constructions, and the related protective effects of the forests. Using a causal model they search for the risks which the forest exerts on other properties.

	forest authority	**torrent and avalanche control service (TACS)**
general field of competence	the forest	natural hazards protection
legal background for activity	Forest Act	Forest Act, Act for Torrents Control, Act for Subsidising Hydrological Engineering, Disaster Relief Fund
object of activity	"protection forests"	"protective forests"
character of projects	protection forest management	watershed management
belief system	doctrines of "forest functions", "wake theory", "close-to-nature management"	doctrine of the "forest-technical system of torrent and avalanche control"
image of ideal forest	timber forest	protective forest
decision model	multifunctional/ vague/ holistic	monofunctional/ specific/ causalistic
goals in mountain forest management projects	combine production and protection goals	maximise protective effects
projected and subsidised measures	general forest management measures	deriving from protection goals
work carried out by:	land owner	TACS workers
project area	extensive	limited
specific costs per ha	low	high
basic perception of risk	risks for forests	risks from forests for properties below
perceived source of risks	elementary forces disturbing the forest	elementary forces, topography and vegetation cover (e.g. forests)
problem definition	risk of forests losing potential to carry out "protection functions"	dangerous processes threatening settlements or infrastructures
protection concept	undifferentiated ("protection functions")	specific ("protective functions")

Table 2. Institutional background, risk perception and management concepts of two agencies active in mountain forest restoration

Discussion and conclusions

In land management it is a typical problem that decisions of certain actors may produce risks for other parties. As the effects of land management for the safety of other parties are often characterised as external effects and public goods, a problem solution strategy using private instruments, e.g. contracts, is difficult and regulations are necessary. The management of forests, for instance, may have negative effects on properties below the forest; the management of watersheds has effects on all properties situated along the river. In the first case the circle of beneficiaries is small which makes it basically possible for them to organise their interests themselves; in the latter a big group benefits from appropriate land management strategies which makes it necessary for the State to intervene, e.g. by taking preventive measures. These public activities again may result in the need for accompanying regulations: As the State covers the costs for hazard prevention and disaster relief, the development of hazard-prone areas for housing etc. has to be restricted. As the property owners' liability for their decisions is limited, they would take higher risks at the cost of others (e.g. hotel guests) or the public (the erection of prevention constructions is subsidised). Furthermore, as areas for development are officially zoned the State also takes responsibility for their safety.

As can be seen, there are several reasons for the State to become active in natural hazard protection. How do public agencies deal with this task? The examples quoted confirm the statement that the perception of risks is strongly determined by the interests and values of the political actors involved in policy-making. The two agencies have different approaches to their work: While the forest authority strives for multifunctional forest management, the TACS applies a monofunctional concept in managing protective forests. General findings about the behaviour of public administration are true for risk regulation as well: Public agencies do not only pursue formal but also informal goals. They are oriented at their main clientele; they want to safeguard their autonomy, territory and legitimacy; they strive for safe decisions; they apply their routine and change policy only under high pressure. Different combinations of main competence field and clientele of agencies on the one hand and certain tasks of risk management on the other produce different problem solving strategies.

One leading principle of bureaucratic behaviour is the need for safe decisions. In the case of the forest authority it was sufficient to define the areas of protection forests with need for restoration only roughly. The criterion of "protection forest" is laid down in the Forest Act, the main legal basis for their activity. The criterion "need for restoration" was assessed by silvicultural expertise and in three priority levels. The assessment was practically desired at the models of the traditional "timber forest" and "close-to-nature" management; for "protective forests", no specific model was developed.

The TACS defined watershed areas and the protective functions of the forest to legitimise their interventions because their principal field of activity is the protection of settlements and infrastructures against torrents and avalanches. In their main field of activity, which is directly connected with the physical processes of natural hazards, the TACS developed more sophisticated risk assessment models. For assessing hazards they introduced the concept of the "probability of recurrence" describing the frequency of certain dangerous processes at a certain location. Their assessments are based upon a standard event with a certain frequency (usually 100 or 150 years). Although the available data are so limited that a sound figure for this frequency cannot be given in most of the cases they still work with this concept. In their hazard zones maps, where they are forced to decide on a certain definition of the area potentially affected by a standard event, the weakness of their database is especially obvious. The map may be questioned by people whose properties are zoned as "not for development" land. For this reason the procedures demand that the hazard zones maps are to be approved by a multi-step evaluation within the TACS hierarchy. The official hazard zones plan is approved by the federal minister of agriculture. The database for the assessment is not improved by this procedure but the decision becomes "safer" symbolically.

Another important informal behaviour of the administration is their orientation at their main clientele. The forest authority's approach is determined by their "multifunctional" forest management concept and their orientation towards forest owners' interests. The multifunctional concept is for a large part a symbolic rhetoric which is used to safeguard the agency's legitimacy because it considers the many good effects of forests for the public. In practice, their policy is strongly oriented at timber production which is generally the main income for forest owners. Other interests, e.g. nature protection concerns, are not pursued unless NGO's exercise high pressure and/or the forest owners are interested themselves. This is the case with community forests which, for instance, are typical of neighbouring Switzerland. Hunting, which is often in the interest of the owner, is officially not supported by the forest authority because this is the responsibility of a different agency. The interventions of the Austrian forest authority in protection forest management were caused by their attempt to safeguard or extend their competence field, their resources and their legitimacy. As the necessity of forest restoration measures was not demanded by the benefiting parties and the agency used public money, they were not forced to develop specific management concepts to meet the needs of this group. The beneficiaries of the protective forests were not included in the policy network. Therefore, the measures are actually for the benefit of the forest owners instead of the property owners whose land is situated below the forest.

The opposite applies to measures carried out by the TACS. They implement measures demanded by property owners/municipalities threatened by hazards. Additionally projects are co-financed by state subsidies and the money of the beneficiaries of the protection measures. However, as these groups are usually more interested in immediate and obvious protection, the TACS is more active in building technical constructions than maintaining the forests. The result of the TACS' orientation at the interests of their clientele is that most of their measures are more for the benefit of economic development than for reducing hazards in general. The hazard zones plans which follow the precautionary principle but restrict land-use are implemented by offering subsidies for preventive constructions in exchange. Zoning hazard-prone areas is accepted locally because the TACS promises subsidised measures, which reduce the hazard as well as the hazard zones. Remaining risks and uncertainties are neglected in the design of constructions and zoning. Defining areas that are protected by technical constructions as free for development leads to future damage or even deaths because of the remaining risks.

The most important lesson to learn from these examples is that the risk policies applied are strongly determined by the stakeholders' interests included in the policy network. Different institutional settings lead to different policies. Experts do not necessarily disagree with these policies; on the contrary, official experts develop different risk concepts and thus different solutions depending on the institutional framework.

On the basis of this insight improvements of the political-administrative system may be found by institutional changes. If certain groups shall benefit from the measures, their inclusion by political participation and/or financial involvement promises more effective and efficient implementation. With these groups participating in decision-making, goals might be formulated more specifically; if benefiting groups are additionally involved in financing, cost-benefit relations might be considered more seriously. At the moment, only powerful groups pursue their interests in the implementation process in an informal way. Formal participation of all groups that are affected by policies would make the political practice of public administrations more democratic. Further research is suggested on ways how beneficiaries and the public could be included in policy-making and on the possible effects of such institutional and procedural innovations.

References

Aulitzky, H., 1968a: Analyse der Schadensursachen von Unwetterkatastrophen zum Zweck der Vorbeugung. *Österreichische Wasserwirtschaft* 20, H. 5/6 und 7/8.

Aulitzky, H., 1970: Der Enterbach (Inzing in Tirol) am 26. 7. 1969: Versuch einer Analyse eines Murganges als Grundlage für die Neuerstellung einer zerstörten Wildbachverbauung. *Wildbach- und Lawinenverbau* 34, H. 1: p. 31-66.

Aulitzky, H., 1996: Beispiele vermeidbarer Hochwasser- und Lawinenschäden. In: Naturkatastrophen – hausgemacht? Conference Proceedings, Österreichische Gesellschaft für Ökologie, Wien: p. 25-55.

Bechmann, G. (ed.), 1993: *Risiko und Gesellschaft. Grundlagen und Ergebnisse interdisziplinärer Risikoforschung.* Westdeutscher Verlag, Opladen.

Beck, U., 1988: *Gegengifte. Die organisierte Unverantwortlichkeit.* Suhrkamp, Frankfurt/Main.

BMLF (Bundesministerium für Land- und Forstwirtschaft), 1978: *Kosten-Nutzen-Untersuchung.* Wien.

BMLF (Bundesministerium für Land- und Forstwirtschaft), 1980: *Richtlinien Hinderungsgründe.* Wien.

BMLF (Bundesministerium für Land- und Forstwirtschaft), 1993: *Projekt Schutzwaldverbesserung.* Wien.

Compare, Alexander, 1995: *Risikokommunikation als Machtproblem.* Europäische Hochschulschriften 22. Lang, Frankfurt/Main.

Conrad, J., 1989: Risiko, Ritual und Politik. In Schmidt, M. (ed.), *Leben in der Risikogesellschaft. Der Umgang mit modernen Zivilisationsrisiken.* Müller, Karlsruhe: p. 179-204.

Dieterich, V., 1953: *Fostwirtschaftspolitik.* Hamburg – Berlin.

Dietz, T.M. and Rycroft, R.W., 1987: *The Risk Professionals.* Sage, New York.

Douglas, M. and Wildavsky A., 1982: *Risk and Culture. An Essay on the Selection of Technological and Environmental Dangers.* University of California Press, Berkeley.

Downs, A., 1967: *Inside Bureaucracy.* Little and Brown, Boston.

Edelmann, M., 1976: *Politik als Ritual.* Deutsche Übersetzung. Campus, Frankfurt/Main.

Elias, N., 1976: *Über den Prozeß der Zivilisation.* 2 Bd., Suhrkamp, Frankfurt/Main.

Evers, A. and Novotny, H., 1987: *Über den Umgang mit Unsicherheit. Die Entdeckung der Gestaltbarkeit von Gesellschaft.* Suhrkamp, Frankfurt/Main.

Frey, W., Forster, B., Gerber, W., Graf, F., Heininger, U., Kuhn, N. and Thee, P., 1995: Risiken und Naturgefahren auf Windwurfflächen. *Schweizer Zeitschrift für Forstwesen* 146 (11): p. 863-872.

Fritsch, M., Wein, T. and Ewers, H.-J., 1993: *Marktversagen und Wirtschaftspolitik: mikroökonomische Grundlagen staatlichen Handelns.* Vahlen, München.

Glück, P., 1976: *Die Rolle der Verbände in der theoretischen Forst- und Holzwirtschaftspolitik.* Habilitationsschrift an der Universität für Bodenkultur, Wien.

Glück, P. and Pleschberger W., 1982: Das Harmoniedenken in der Forstpolitik. *Allgemeine Forstzeitschrift* 37 (22): p. 650-655.

Glück, P. and Weber, M. (eds.), 1998: *Mountain Forestry in Europe – Evaluation of Silvicultural and Policy Means.* Publication Series of the Institute for Forest Sector Policy and Economics 35. Universität für Bodenkultur Wien.

Glück, P. and Weiss, G., 1997: Evaluierung der Instrumente zur Sicherung der Schutzleistung des Waldes – Umweltpolitische Prinzipien in der Implementierung. In BUWAL, *Ökosystem Wald - Rohstoff Holz - Prinzip Nachhaltigkeit.* Umweltmaterialien Nr. 67. Bundesamt für Umwelt, Wald und Landschaft, Bern, p. 45-56.

Heeg, B., 1973: Forstliche Funktionenlehre und forstpolitische Theorie. In: *Allgemeine Forst- und Jagdzeitschrift* 144 (12): p. 242-251.

Héritier, A. (ed.), 1993: *Policy-Analyse. Kritik und Neuorientierung.* Politische Vierteljahresschrift (PVS), Sonderband 24/1993. Westdeutscher Verlag, Opladen.

Hollenstein, K., 1995: Analyse und Bewertung von Risiko und Sicherheit bei Naturgefahren. In: *Schweizer Zeitschrift für Forstwesen* 146/9: p. 687-700.

Jäggi, M., 1996: Gedanken zu den Grenzen der Vorbeugung. *Österreichische Wasser- und Abfallwirtschaft* 48, H. 5/6: p. 179-186.

Jänicke, M., 1986: *Staatsversagen: die Ohnmacht der Politik in der Industriegesellschaft.* Piper, München, Zürich.

Johnson, B.B. and Covello, V.T., 1987: *The Social and Cultural Construction of Risk.* Reidel, Dordrecht.

Jungermann, H. and Slovic, P., 1993: Die Psychologie der Kognition und Evaluation von Risiko. In: Bechmann (ed.): p. 167-207.

Kaufmann, F.-X. (ed.), 1991: *The Public Sector. Challenge for Coordination and Learning.* deGruyter, Berlin, New York.

Kienholz, H., 1993: Naturgefahren – Naturrisiken im Gebirge. *Forum für Wissen*: p. 7-22.

Koch, N.E. and Rasmussen, J.N. (eds.), 1998: *Forestry in the Context of Rural Development. Final Report of COST Action E3.* Horsholm: Danish Forest and Landscape Research Institute.

Krohn, G. and Krücken, W., 1993: Risiko als Konstruktion und Wirklichkeit. Eine Einführung in die sozialwissenschaftliche Risikoforschung. In Krohn/Krücken (eds.): p. 9-44.

Krohn, G. and Krücken, W. (eds.), 1993: *Riskante Technologien: Reflexion und Regulation. Einführung in die sozialwissenschaftliche Risikoforschung.* Suhrkamp, Frankfurt/Main.

Krott, M., 1985: Zu den Waldfunktionen als Instrument der forstpolitischen Wissenschaft und Praxis. *Centralblatt für das gesamte Forstwesen* 102 (1): p. 1-28.

Krott, M., 1990: *Öffentliche Verwaltung im Umweltschutz. Ergebnisse einer behördenorientierten Policy-Analyse am Beispiel Waldschutz.* Braumüller, Wien.

Lowrance, W.W., 1980: The nature of risk. In Schwing/Albers (eds): p. 5-14.

Luhmann, N., 1966: Politische Planung. *Jahrbuch für Sozialwissenschaft.* Göttingen.

Luhmann, N., 1993: Risiko und Gefahr. In: Krohn/Krücken (eds.): p. 138-185.

Mayer, H., 1967a: Das Fichten-Naturwaldreservat Rauterriegel am Eisenhut bei Turrach (Österreich). *Centralblatt für das gesamte Forstwesen* 84: p. 2-6.

Mayer, H., 1967b: Zur Behandlung überalterter Gebirgswälder. In: *Schweizer Zeitschrift für Forstwesen* 118 (6): p. 335-372.

Mayer, H., 1976: *Gebirgswaldbau-Schutzwaldpflege.* Stuttgart.

Mayntz, Renate, 1985: *Soziologie der öffentlichen Verwaltung.* 3., überarb. Auflage. Müller, Heidelberg.

Mazur, A., 1985: Bias in Risk-Benefit Analysis. *Technology in Society* 7: p. 25-30.

Meister, G., 1985: Übergangsstrategie zur Sicherung der Schutzfunktionen des Gebirgswaldes. *Mitteilungen des Deutschen Alpenvereins,* H. 3: p. 149-155.

Müller, W., 1993: Risiko und Ungewißheit. *Enzyklopädie der Betriebswirtschaft:* p. 3813-3825.

Nelkin, D. and Pollak M., 1980: Problems and Procedures in the Regulation of Technological Risk. In: Schwing/Albers (eds.): p. 233-248.

Olson, M., 1965: The Logic of Collective Action. Cambrigde (Mass).

Ottitsch, A. and Weiss, G., 1998: Mountain Forest Policies in European Countries – a Comparison Using RAGIN's Qualitative Comparative Analysis Method. In Glück/Weber (eds.): p. 263-284.

Ottitsch, A., Pregernig, M. and Rappold, G., 1998: *Bürgerbeteiligung in der forsttechnischen Planung.* Research report prepared for the Bundesministerium für Land- und Forstwirtschaft, Wien.

Perrow, C., 1984: Normal Accidents. Living with High-Risk Technologies. Basic Books, New York. Dt.: Normale Katastrophen. Die unvermeidbaren Risiken der Großtechnik. Campus, Frankfurt/Main and New York.

Peters, B.G., 1984: *The Politics of Bureaucracy.* New York and London.

Pleschberger, W., 1989: *Staat und Wirtschaft. Am Beispiel der österreichischen Forstgesetzgebung von 1950 bis 1987.* Studien zu Politik und Verwaltung Bd. 28. Böhlau, Wien and Köln.

Pregernig, M. and Weiss G., 1998: *Forest Policy in Austria: Policy Making by the Sector for the Sector.* Paper presented at the IFF (Intergovernmental Forum on Forests) European Regional Meeting on the Underlying Causes of Deforestation and Forest Degradation in Europe, 28-29 October, Bonn.

Prittwitz, V., 1990: *Das Katastrophenparadox. Elemente einer Theorie der Umweltpolitik.* Leske+Budrich, Opladen.

Rosa, E.A., 1997: *Modern Theories of Society and the Environment: The Risk Society.* Paper presented at the Conference „Sociological Theory and the Environment", 20-23 March 1997, Woudschoten, The Netherlands.

Sabatier, P.A., 1986: The Top-Down and Bottom-Up Approaches to Implementation Research: a Critical Analysis and Suggested Synthesis. *Journal of Public Policy* 6(1) 21-48.

Sabatier, P.A., 1993: Advocacy-Koalitionen, Policy-Wandel und Policy-Lernen: Eine Alternative zur Phasenheuristik. In: Héritier (ed.): p. 116-148.

SBN, 1994: *Naturschutz im Bergwald. Grundlagenpapier in Zusammenarbeit mit der Schweizerischen Gebirgswaldpflegegruppe.* Schweizerischer Bund für Naturschutz, Basel.

Scharpf, F. W., 1997: *Games Real Actors Play. Actor-Centered Institutionalism in Policy Research.* Westview Press, Boulder, Oxford.

Schmiderer, H. and Weiss, G., 1999: *Die Bannwaldpolitik in Österreich.* Publication Series of the Institute for Forest Sector Policy and Economics 38. Universität für Bodenkultur, Wien.

Schwing, R.C. and Albers, W.A. (eds.), 1980: *Societal Risk Assessment. How Safe is Safe Enough?* Plenum Press, New York.

Shrader-Frechette, K.S., 1995: Evaluating the expertise of experts. *Risk* 6(1).

Shrader-Frechette, K.S., 1997: How some risk frameworks disenfranchise the public. *Risk* 8(4).

Sieferle, R.P., 1997: *Rückblick auf die Natur: eine Geschichte des Menschen und seiner Umwelt.* Luchterhand, München.

Starr, C., 1969: Societal Benefit versus Technological Risk. What is Our Society Willing to Pay for Safety? *Science* 165: p. 1232-1238.

Starr, C., 1980: Introduction to the session. In Schwing/Albers (eds): p. 1-4.

Thompson, P.B. and Dean, W., 1996: Competing conceptions of risk. *Risk* 7(3).

Weinmeister, H.W., 1988: Ökonomie und Ökologie am Beispiel des Schutzwasserbaues (landschaftsökologische und volkswirtschaftliche Grenzen des technischen Schutzwasserbaus). Proceedings of the conference *Interpraevent* 1988, vol. 4: p. 339-376.

Weiss, G., 1998a: *Bewertung der Maßnahmen zur Schutzwaldverbesserung.* Research Report prepared for the Bundesministerium für Land- und Forstwirtschaft, Wien, 4 volumes.

Weiss, G., 1998b: Evaluation of Mountain Forest Policy in Austria. In Glück/Weber (eds.): p. 107-136.

Weiss, G., 1999: *Die Schutzwaldpolitik in Österreich. Einsatz forstpolitischer Instrumente zum Schutz vor Naturgefahren.* Proceedings of the Institute of Forest Sector Policy and Economics 39, Wien. Also: dissertation at the Universität für Bodenkultur, Wien.

Weiss, G. (in press) Perceptions of risks in mountain forests. In M. Price (ed.), *State-of-knowledge-report on forests in mountain development.* IUFRO/FAO.

Wildavsky, A., 1989: *Searching for safety.* Oxford (USA).

Wolf, R., 1986: *Der Stand der Technik. Geschichte, Strukturelemente und Funktion der Verrechtlichung technischer Risiken am Beispiel des Immissionsschutzes.* Westdeutscher Verlag, Opladen.

CONSIDERING SERIOUS HAZARDS IN FOREST MANAGEMENT DECISION-MAKING

Matthias Dieter [1], *Martin Moog* [2], *Herbert Borchert* [3]

Abstract

A forest enterprise manager runs risks, when deciding about harvesting strategies. The postponement of the harvest of a stand may yield, with a specificprobability, a higher profit. However, as a consequence of storm damage, financial losses may occur. The risk of storm damage varies between different stands, similar to the price risk between shares. In this study, the problem of stand selection for final cuts is compared with the problem of optimising a portfolio. Investors that consider only the expected net present value in their investment selection are regarded as risk neutral. Empirical studies show that investors who also consider the range of the profit dispersion are usually risk averse. As a subjective decision criterion from the field of portfolio selection, the term "expected utility" is used. The expected utility results from reduction of the expected value subject to the level of risk (variance) and the risk aversion of the decision maker. When the decision involves more investments, the risk correlation between the different investments is also taken into account. By means of examples, the use of this decision criterion in forest management is shown. One example concerns the selection of harvesting strategies. Considering a number of equally mature stands, a forest manager has to decide which stands to cut either today, or 20 or 40 years later. Waiting yields a higher profit, if no storm damage occurs. In case of a damage, the income is less than the present stumpage value. In the second example, different probabilities of storm damage are allowed for the stands. The optimal harvesting strategy can be selected using a non-linear optimisation method. Although currently suitable decision support models from the field of Operation Research exist, formal and conscious risk management has not been implemented in German forestry. Finally, the obstacles for the implementation of improved planning tools in forest management are discussed and the future prospects of their application are assessed.

Keywords: *Risk management, portfolio selection, expected utility, risk correlation, storm damage, harvesting strategy, risk attitude, non-linear optimisation.*

[1] Dr. Matthias Dieter, Institut für Ökonomie der Bundesforschungsanstalt für Forst- und Holzwirtschaft, Leuschnerstraße 91, 21031 Hamburg

[2] Prof. Dr. Martin Moog, Lehrstuhl für Forstl. Wirtschaftslehre der Forstwissenschaftlichen Fakultät, Technische Universität München, Am Hochanger 13, 85354 Freising

[3] Herbert Borchert, Lehrstuhl für Forstliche Wirtschaftslehre der Forstwissenschaftlichen Fakultät, Technische Universität München, Am Hochanger 13, 85354 Freising

K. von Gadow (ed.), Risk Analysis in Forest Management, 201–240.

1 Importance of risk in forest management

The effects of the 1990 spring storms, which caused extensive windthrow in Central Europe, wood accumulations, consequential Bark beetle calamities and a price crash as a result of the additional supply of storm timber, have sensitised decision makers in German forest enterprises to risk in forest management. Subsequently, efforts in forest research have been intensified to increase the knowledge available on the impact of catastrophic risks and thus contribute to the improvement of decision making in forest enterprises. In research projects concerning the so-called "Waldsterben", aspects of risk have already been discussed by German forest economists (Möhring, 1986). However, since in traditional forest valuation risk is almost completely ignored, as it is in silvicultural models, this can be regarded as a new field of research in forest economics.

The extent and the distribution of storm damage was recorded in several studies published during the 1990's (König et al, 1995; Richter, 1996; Waldherr, 1997; Richter, 1998), whereby the basic data for models that explain storm damage were improved and a model was developed for forest regions in Bavaria (König, 1995, 1996). The most important site characteristic input variables in this model are (with decreasing importance), the mean height of dominant trees, site quality, the time delay from the last thinning and the maximum wind velocity (König, 1995, p. 65).

With the quantification of the most influential factors on storm damage, preconditions for the inclusion of the risk of storm damage in silvicultural and forest enterprise decisions can be satisfied. Using this model as an empirical basis, models for different forest enterprise decisions can be designed, both on an individual forest stand level and on an operational level (Dieter, 1997). However, further improvement of this empirical basis seems to be necessary because so far no empirical information about the attitude of decision makers in forest enterprises towards risk is available.

Since the switch from conifer to broad-leave forest planting, which started in the 80s and increased after the 1990 storms, one of the most essential questions is whether mature conifer stands should be underplanted with broad-leaves or whether they should be reforested after a clearcut.

Before young broadleaves can be planted under old conifers, the stand must usually be opened. In comparison with a plantation after a clearcut, this silvicultural technique requires lower expenses. Additionally, the income from the trees harvested to open the stand is received earlier, however, this is also true for the expenses. Whether or not this silvicultural technique is a good choice depends on the risk of storm damage. Since these stands are more or less unstable for some years after the thinning, the risk increases and may cause substantial losses (Dieter, 1997, 1999).

Also important for the economic success of forest enterprises is the planning of final cuttings. Considering the risk of catastrophic events, rotation periods obviously should be planned separately for different sites. While the optimisation of the rotation for a single stand as well as of an average rotation period for all stands of a forest enterprise are problems often discussed, the real situation of decisionmaking in a forest enterprise may be better described by the following problem formulation:

> Select the optimal combination of stands for harvesting from all mature stands, under the constraint that a certain volume of timber must be cut and a certain volume must remain.

This problem formulation makes clear that the risk of storm damage should be considered if it is significantly different for each stand. Stands with a relatively low risk can be set aside for use in future periods while stands with a high risk may be cut relatively young in order to prevent future losses.

The parallelis of this problem to the optimisation of a portfolio is obvious. Therefore, the problem of stand selection for final cuts will be discussed from this perspective. In keeping with this vision, some basics of decision-making and of investment selection will be presented briefly in the following section.

2 Selection of investments - risk averse and risk seeking investors

The first prerequisite for a rational investment is a rating of profitability. A suitable parameter is the expected net present value (or the final value, final net worth or the net yield under certain conditions), which can be a measure for gain alone or its variance can be used as a measure of risk.

Different investments can be represented by points on a coordinate system with the expected value of the net present value μ on the abscissa and the standard deviation σ (root of the variance) on the ordinate. With such a μ / σ -diagram, different profit-risk-relations can be shown and in certain circumstances the superiority of an investment can easily be visualised (see fig. 1).

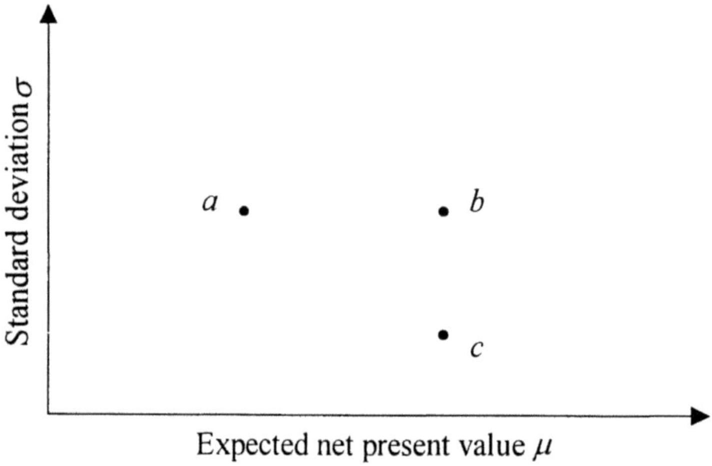

Figure 1. Investments a, b and c with different expected net present values, μ, and different standard deviations, σ, of the possible net present values - situation of superiority of one investment.

A rational investor will prefer a project with a higher net yield as long as it represents less or equal risk. In figure 1, the investment characterised by point *b* has a higher net yield than the investment characterised by point *a*, while the risk is the same. The investment characterised by point *c* in turn has the same net yield as point *b*, but a smaller risk. In this case, a selection between the investments *a*, *b* and *c* is possible even without considering the investor's risk preference.

Sometimes an investor must choose between two investments in a situation without a clear superiority of one investment. It is not uncommon that one investment is characterised by a higher net yield and a higher risk than the other investments. If an investor must decide between these investments, the question arises whether the higher risk is compensated by the higher net yield (see fig. 2).

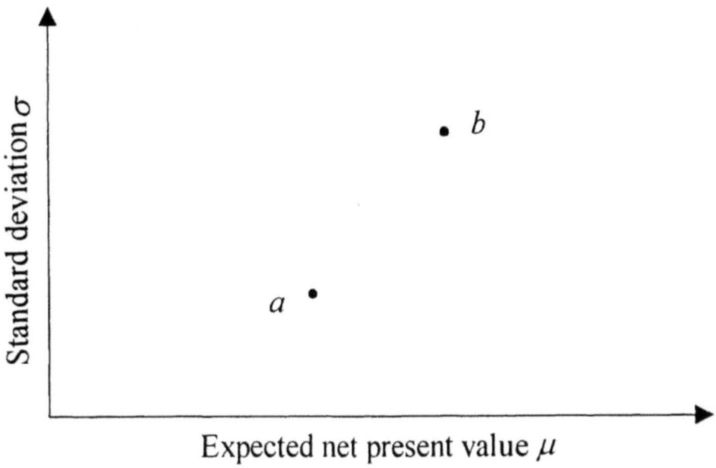

Figure 2. Investment projects with different expected net present values, μ, and different standard deviations, σ, of the possible net present values – no clear superiority of an alternative.

Investors in such situations that follow the so called μ - principle decide without considering different levels of risk and will be regarded as risk neutral. They make decisions without considering the variance or standard deviation of the profits and will choose the investment with the higher profit. In other words, risk neutral investors would pay up to the expected net present value for an investment with uncertain output. This variable is called safety equivalent, because it is the safe amount which is valued equal to the uncertain result of the investment by the investor. If investors make their decisions following this principle and thus select the investment with the higher expected value, then the selection doesn't consider the dispersion of the results.

Empirical studies of decision behaviour show that investors do not select investments based solely on their expected values but also on the range of the profit dispersion. They can be regarded either as risk-averse or risk-seeking. A risk-averse investor would not pay the amount of the expected net present value for an investment with uncertain profits. His safety equivalent for this project is smaller than the expected net present value. On the other hand a risk-seeking investor is willing to pay more than the expected net present value for an uncertain investment. For a risk-seeking investor, the safety equivalent of the probability distribution of the profit of the project is higher than the expected net present value (Blohm and Lüder, 1991, p. 242; Kruschwitz, 1995, p. 256). Risk-averse investors select projects with higher profits and higher risk only in cases where they are overcompensated with higher profits. In order to accept a higher risk, they demand a higher expected value of the profit. They are indifferent to investment projects that lie on the concave curves of the □ / □ - coordinate system. In the situation presented in figure 3, the investor would prefer the project *a* to the project *b*, because it lies on a higher indifference curve. The more significant the risk aversion, the stronger the indifference lines are curved.

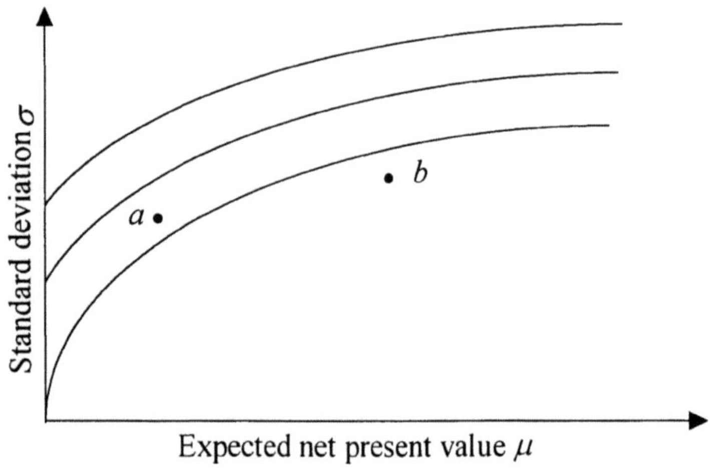

Figure 3. Investment projects a and b in a μ / σ - diagram and indifference curves of a risk averse investor. The system of indifference curves characterise the utility function of the investor.

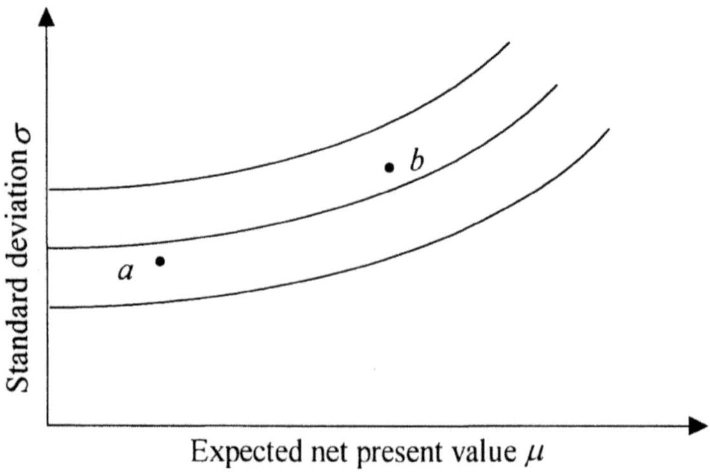

Figure 4. Investment projects a and b in a μ / σ - diagram and indifference curves of a risk seeking investor.

Risk-lovers prefer the investments with broader standard deviation, if the expected net present values are equal. The indifference curves of risk-seeking investors are therefore curved convex to zero point of the μ / σ -coordinate system. In the situation presented in figure 4, the risk-seeking investor would prefer project *b*, because it is now situated on a higher indifference curve.

Empirical investigations have given evidence that investors are usually risk-averse when selecting investments.

3 The effect of risk correlation

The total risk, that an investor takes by buying or keeping an investment portfolio, depends on the risk correlation between the individual investments. This is a widely used principle in risk control, particularly in financial markets. A forest enterprise can be regarded as a risky investment portfolio as well. In forest enterprise decisions, especially concerning final cuttings, the question is whether a risk correlation exists between individual stands and if they have an important effect on the risk of the forest enterprise as a whole.

The variance of the market values of a share portfolio helps the investor to assess the risk of his investment. Thus, not only the variances of the prices of the individual shares are of importance, but also their covariance. Obviously, the total risk of two investments also depends on the probability that changes in their values (share market price) are parallel or inverse.

The variance of a discrete random variable *X*, e.g. the market value of a share portfolio or the random income (stumpage value) from a forest stand with high storm risk at a particular time, can be calculated using formula (1):

$$Var(X) = \sum_{i=1}^{n} f_X(x_i) \cdot (x_i - E(X))^2 \qquad (1)$$

where $f_X(x_i)$ is the probability that the random variable X has the value x_i at time $i = 1 \ldots n$, and $E(X)$ is the expected value for X.

The variance of the total random stumpage values X_1, X_2 X_3 ... X_k from stands $j = 1 \ldots n$ where storm damage may occur, can be calculated with the following formula, where a is the proportion of each forest stand in the total area:

$$Var(\sum_{j=1}^{n} a_j \cdot X_j) = \sum a_j^2 \cdot Var(X_j) + \sum_{k \neq m} a_k \cdot a_m \cdot Cov(X_k, X_m) \quad (2)$$

The quotient of covariance Cov, and the standard deviations σ, of the individual random stumpage values equals the coefficient of correlation ρ, which is of immense importance for the dispersion of the total income from two different stands:

$$\rho = \frac{Cov(X_k, X_m)}{\sigma_k \cdot \sigma_m} \quad (3)$$

The coefficient of correlation ρ, has values which range between 1 and -1. The closer its value is to 1 or -1, the more probable is a relationship between the random events, which would be storm damage in this example. A coefficient of correlation of 0 means that storm damage in one stand is unrelated to that of another stand.

In the case of a strong negative coefficient of correlation, an inverse relationship exists. That is to say, if the value of one investment decreases the value of the other investment increases. In the case of an investment portfolio, this would mean that a loss caused by the decrease of the price of one investment is compensated with profit from the increase of the price of the other investment. In case of a perfect negative correlation, the risk measured by the common dispersion disappears. This effect can be demonstrated with the following simple example (see Spremann, 1991, p. 456).

Two investments with the profits R_1 and R_2 have the expected values $\mu = E(R_1) = E(R_2)$ and the variances $Var(R_1) = Var(R_2) = \sigma^2$. The expected value and variance of profit of the portfolio R_M can be calculated with the following formulas. To make it simple, we assume that fifty percent of the financial resources are allocated to each investment.

$$R_M = \frac{1}{2} \cdot R_1 + \frac{1}{2} \cdot R_2 \quad (4)$$

$$E(R_M) = \frac{1}{2} \cdot E(R_1) + \frac{1}{2} \cdot E(R_2) = \mu \quad (5)$$

$$Var(R_M) = \frac{1}{4} \cdot Var(R_1) + \frac{1}{4} \cdot Var(R_2) + \frac{1}{2} \cdot Cov(R_1, R_2) \quad (6)$$

$$= \frac{1}{2} \cdot \sigma^2 + \frac{1}{2} \cdot \sqrt{Var(R_1)} \cdot \sqrt{Var(R_2)} \cdot \rho \quad (7)$$

For $Var(R_1) = Var(R_2) = \sigma^2$ it applies:

$$Var(R_M) = \frac{1}{2} \cdot \sigma^2 + \frac{1}{2} \cdot \sigma^2 \cdot \rho = \frac{1}{2} \cdot \sigma^2 \cdot (1 + \rho) = \frac{1 + \rho}{2} \cdot \sigma^2 \quad (8)$$

In case capital is allocated to two investments with equal expected values and the same variance, the expected value of the combination of the investments is equal to those of the individual investments. However, the dispersion measured by the variance of the combination investment depends substantially on the correlation of the two investments.

When the coefficient of correlation is $\rho = 1$, the variance of the combination of the investments R_M is equal to the variances of the individual investments R_1 and R_2. If there is no correlation between the two random profits from the two investments, the variance of the combination of the investments R_M is only half the amount of the variance of R_1 and R_2. In the case of a negative correlation between the profits of R_1 and R_2, then for $\rho = -1$, the risk of the combination of the investments is zero. It is obvious how important it is to have information about the correlations of the random net yields when deciding on risky investments.

4 The correlation of storm damage

As shown above, the correlations of random profits from several investments are important for the selection of an optimal combination of investments. This idea can be transferred to forest enterprise management. In order to enable a decision maker with a specific risk tolerance to choose an optimal stand composition, it is necessary to know the coefficients of correlation of storm damage for the different stands.

When only the stumpage values at different ages and their relevant probabilities for individual stands are known, it is necessary to calculate a common probability distribution for two stands, which in turn can be used to calculate the covariance and thus the coefficient of correlation.

To simplify the problem, we assume that not only the probabilities of damage at a certain period are known, but also the probabilities for the survival of a stand during that period. Furthermore, we will consider only two stands and two intervals. This example is used to demonstrate how a common probability distribution can be derived.

The probability that stand A is damaged in x^{th} interval by a storm is a_x. Similarly, the probability that stand B is damaged in x^{th} interval is b_x. The probability for stand A to survive two intervals is $(1 - a_1) \cdot (1 - a_2)$ and for stand B it is $(1 - b_1) \cdot (1 - b_2)$.

An empirical study (König, 1995) indicates that the probability a_X for storm damages in a stand A is determined by the probability of the occurrence of a storm and by factors such as height of the stand, site characteristics and time lag since the last thinning, which influence the probability of storm damage.

The known probability a_x is therefore the product of the probability for a storm StW and the stand-specific probability for damage a_{xsp}. It follows that: $StW \cdot a_{xsp} = a_x$. A stand survives the interval x, if either no storm occurs ($1 - StW$) or no damage occurs ($1 - a_{xsp}$). Similarly, for B: $StW \cdot b_{xsp} = b_x$.

The probabilities a_x and b_x consist of an independent term a_{xsp} and b_{xsp} and of a uniform storm probability StW. Thus, the desired common probability distribution can be calculated. This distribution aims at designing a matrix for stands A and B, in which common probabilities for achieving all possible combinations of the stand developments of A and B are presented. Thus the following situations can be distinguished (see table 1):

$a_1 =$ stand A is damaged during the first interval,

$a_2 =$ stand A survives the first interval and is damaged during the second interval and

$g_2 = 1 - a_2 =$ stand A survives the first and the second interval.

Likewise:

$b_1 =$ stand B is damaged during the first interval,

$b_2 =$ stand B survives the first interval and is damaged during the second interval and

$h_2 = 1 - b_2 =$ stand B survives the first and the second interval.

Additionally, the theoretical combination possibilities for the conditions of the two stands A and B are shown using a probability tree (see fig. 5). In the analysis of all possible combinations of the stand conditions, depending on the environmental conditions "storm" or "no storm", 17 different paths or sequences of events were found, whose total probabilities were equal to one.

Every path in the tree is related to one field of the matrix in table 1. For example, in sequence 3 there is a storm during the first interval, in which stand A is damaged and stand B survives. During the second interval another storm occurs and stand B survives again. Thus, this sequence belongs to the field of the matrix that indicates a coincidence with the results "stand A is damaged during the first interval, stand B survives both intervals"(last row, first column).

Another example can be seen with sequence 4, where the same result is achieved: a storm occurs in the first interval, stand A is damaged and stand B survives and there is no storm during the second interval. Sequence 4 also belongs to the field of the matrix that has already been described.

Stand A / Stand B	a_1	a_2	g_2
b_1	A is damaged during the first interval and B is damaged during the first interval	A is damaged during the second interval and B is damaged during the first interval	A survives the first and second intervals and B is damaged during the first interval
b_2	A is damaged during the first interval and B is damaged during the second interval	A is damaged during the second interval and B is damaged during the second interval	A survives the first and second intervals and B is damaged during the second interval
h_2	A is damaged during the first interval and B survives the first and second intervals	A is damaged during the second interval and B survives the first and the second interval	A survives the first and second intervals and B survives the first and second intervals

Table 1. Matrix showing all possible combinations of stands A and B development in two time periods during which storm damage may occur.

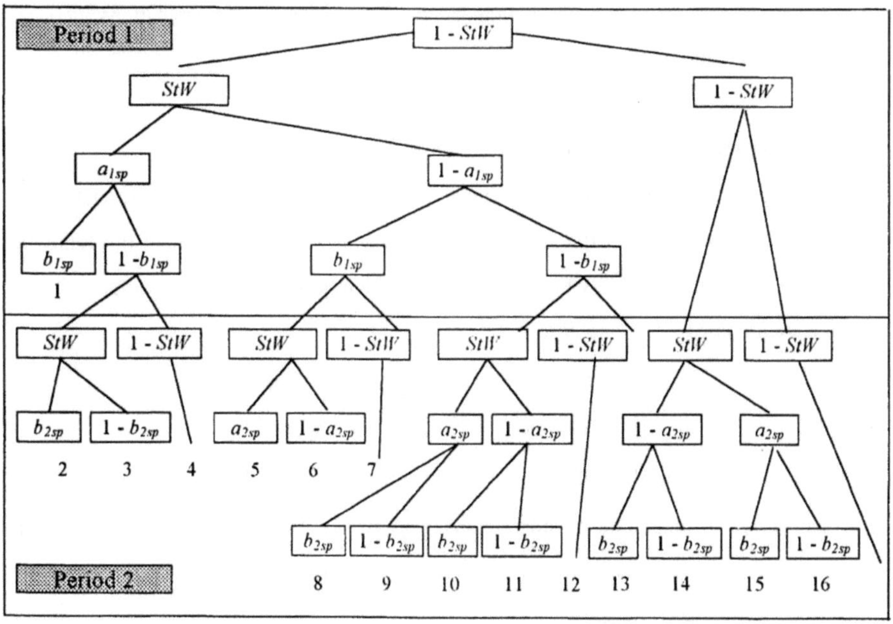

Figure 5. Probability tree for all possible sequences of events in two periods.

The matrix is filled with the paths using this logic (see table 2). The probability of each path and the sum of the probabilities for each field of the matrix can be calculated, whereby the necessary numerical basis for the calculation of covariance and coefficient of correlation of incomes from stands *A* and *B* is created.

		Stand *A* is damaged during the first interval	Stand *A* is damaged during the second interval	Stand *A* survives the first and second intervals
		a_1	a_2	g_2
Stand *B* is damaged during the first interval	b_1	1	5	6 / 7
Stand *B* is damaged during the second interval	b_2	2	8 / 13	10 / 15
Stand *B* survives the first and second intervals	h_2	3 / 4	9 / 14	11 / 12 / 16 / 17

Table 2. Matrix of the probabilities of all possible combinations of the conditions of stand A and B, showing the numbers of the paths, with which the probabilities are to be calculated.

The probability distributions for more intervals can also be calculated using this method. While manual determination of the probabilities by multiplication and addition of values of a probability pyramid is theoretically possible, with larger matrices an EDP (Electronic Data Processing)-supported calculation is recommended, as the number of combinations increases exponentially with increasing number of intervals.

5 Covariance and coefficient of correlation

The probability of a storm damage to a stand during a time interval is the product of the probability of a storm *StW* and the stand-specific probability for a damage. In the following example, a 15 % storm probability in all intervals and the following stand-specific probabilities of damage are assumed:

	Stand *A*	Stand *B*
Interval 1	$a_{1sp} = 0.66666667$	$b_{1sp} = 0.26666667$
Interval 2	$a_{2sp} = 0.86666667$	$b_{2sp} = 0.66666667$

The probability of incomes due to storm damage in stands *A* and *B*, at both intervals, are computed as the product of $StW \cdot a_{isp}$ and $StW \cdot b_{isp}$ respectively:

	Stand *A*	Stand *B*
Interval 1	$a_1 = 0.1$	$b_1 = 0.04$
Interval 2	$a_2 = 0.13$	$b_2 = 0.1$

If stand A is windblown during interval 1 with the probability $a_1 = 0.1$, it will survive this interval obviously with the probability of $1 - a_1 = 0.9$. The conditional probability that it will be blown down during interval 2 is $(1 - a_1) \cdot a_2 = 0.9 \cdot 0.13 = 0.117$. The conditional probability that the stand will survive both intervals is: $(1 - a_1) \cdot (1 - a_2) = 0.9 \cdot 0.87 = 0.783$. Similarly, these probabilities can be calculated for stand B. If the stands survive both intervals, they should be cut according to harvesting plan. The following discounted stumpage values are assumed in the different situations:

	Stand *A*		**Stand *B***	
	probability	stumpage value	probability	stumpage value
storm damage during interval 1	0.100	$ 18,000	0.040	$ 16,000
storm damage during interval 2	0.117	$ 26,000	0.096	$ 24,000
use according to plan	0.783	$ 76,000	0.864	$ 70,000
	1		1	

The probabilities of all combinations can be calculated using the storm probability and the stand-specific damage probabilities (see tab. 3).

	a_1	a_2	g_2
b_1	0.02666667	0.00173333	0.0116
b_2	0.00733333	0.07684444	0.01182222
h_2	0.066	0.03842222	0.75957778

Table 3. Probabilities of all possible combinations of stand development of stands A and B.

As an example, it can be seen on table 2 that the combination a_1 / h_2 is based on the sequences of events 3 and 4 of the probability tree in figure 5. The probabilities of these sequences are calculated as follows:

Sequence 3: $StW - a_{1sp} - (1 - b_{1sp}) - StW - (1 - b_{2sp})$

Sequence 4: $StW - a_{1sp} - (1 - b_{1sp}) - (1 - StW)$

The probability of 0.066 for combination a_1 / h_2 is calculated by adding the probabilities for sequences 3 and 4.

The expected value E, the variance Var, the standard deviation σ, the covariance Cov, and the coefficient of correlation □, can now be calculated.

$E(A) = 0{,}1 \cdot 18.000 + 0{,}117 \cdot 26.000 + 0{,}783 \cdot 76.000 = 64.350$

$E(B) = 0{,}04 \cdot 16.000 + 0{,}096 \cdot 24.000 + 0{,}864 \cdot 70.000 = 63.424$

$Var(A) = 0{,}1 \cdot (18.000\text{-}64.350)^2 + 0{,}117 \cdot (26.000\text{-}64.350)^2 + 0{,}783 \cdot (76.000\text{-}64.350)^2 = 493.177.500$

$Var(B) = 0{,}04 \cdot (16.000\text{-}63.424)^2 + 0{,}096 \cdot (24.000\text{-}63.424)^2 + 0{,}864 \cdot (70.000\text{-}63.424)^2 = 276.532.224$

$\sigma(A) = 22.208$

$\sigma(B) = 16.629$

$Cov(A,B) = 0{,}02666667 \cdot (18.000\text{-}64.350) \cdot (16.000\text{-}63.424) + 0{,}00173333 \cdot (26.000\text{-}64.350) \cdot (16.000\text{-}63.424) + 0{,}0116 \cdot (76.000\text{-}64.350) \cdot (16.000\text{-}63.424) + 0{,}00733333 \cdot (18.000\text{-}64.350) \cdot (24.000\text{-}63.424) + 0{,}07684444 \cdot (26.000\text{-}64.350) \cdot (24.000\text{-}63.424) + 0{,}01182222 \cdot (76.000\text{-}64.350) \cdot (24.000\text{-}63.424) + 0{,}066 \cdot (18.000\text{-}64.350) \cdot (70.000\text{-}63.424) + 0{,}03842222 \cdot (26.000\text{-}64.350) \cdot (70.000\text{-}63.424) + 0{,}75957778 \cdot (76.000\text{-}64.350) \cdot (70.000\text{-}63.424) = 207.897.156$

$\rho = 207.897.156 \,/\, (22.208 \cdot 16.629) \approx 0{,}56$

6 Planning cutting strategies

Using the expected values, variances and coefficients of correlation calculated above an optimal cutting strategy can be defined. The following example illustrates this methodology. However, the personal risk attitude of decision makers must be dealt with separately in greater detail.

6.1 Risk attitude ratios

A characteristic ratio describing risk attitude will be used for the present model. It is the result of statistical analysis of the behaviour of numerous private investors (Spremann, 1991, p. 448).

As a subjective decision criteria, the term "expected utility" $W(x_i)$ will be used, a term that originated from the field of portfolio selection. If x_1 and x_2 are known quantities of the investments 1 and 2 and $V(x_1, x_2)$ is the discounted payment, dependent on final net values, then the expected utility $W(x_i)$ is defined as:

$$W(x_1,x_2) := E[V(x_1,x_2)] - \frac{a}{2}\, Var[V(x_1,x_2)] \tag{9}$$

The expected present net value $E[V(x_1, x_2)]$, represents the mathematical expectation of the investments 1 and 2. The term $a/2$, a being risk aversion and $a \geq 0$, is called proportionality constant (Spremann, 1991, p. 447). Thus, the expected utility is a result of the reduction of a variance proportional constant of personal risk attitude, from expected net value of the investments. This decision support model therefore includes risk through a characteristic ratio and the variance of the random net profit (calculated with the stumpage value when forests are considered).

The risk aversion a was investigated in empirical investigations of numerous private investors (Spremann, 1991, p. 448). For most people it has been proven that a is depending on the amount of money that can be invested, b. It follows that: $a = 1/b$. Risk neutral decision makers assess risky investments using only the expected value criteria. Their expected utility $W(x)$ is equal to the expected value $E[V(x)]$, thus a is zero (see formula 9) for risk neutral investors. The expected utility of risk averse people decreases with increasing risk, while the expected value does not change. The more risk averse an investor is, the stronger the influence of risk on his decision. Consequently, the proportionality constant $a/2$ must have a higher value for especially risk averse investors, for example $a = 2/b$. However, risk aversion a, can not exceed an amount that yields an expected utility smaller than the value of the secure income. Since risk aversion was determined on the basis of formula 9, a can only be used in connection with a risk evaluation based on the expected value and variance.

6.2 Example 1

The fòllowing example of a harvesting strategy will be determined using the expected values, variances and coefficients of correlation calculated previously.

A forest enterprise with a fixed number of equally large mature spruce stands is given. Each stand comprises one hectare and is 80 years old. Since the forest enterprise has 100 of such stands at its disposal, the total area that can be cut amounts to 100 hectares. The risk aversion of the forest owner is characterised by $a = 1.666 \cdot 10^{-7}$. The stumpage value of an 80 year old stand is \$ 60,000. Therefore, the total stumpage value of the area is \$ 60,000 · 100 = \$ 6,000,000 = b. As a result, $a = 1/b$, it follows that for the private investor: $a = 1.666 \cdot 10^{-7}$, which indicates that the forest owner is equally risk averse as the average private investor. Under these circumstances, the forest owner generally has three different cutting strategies:

Cutting strategy 1

All the 80 year old stands can be harvested within a short time. Thus, for each stand he receives a fixed income of \$ 60,000 without any risk.

Cutting strategy 2

The 80 year old stands are not cut but are left to grow for another 20 years, so that they can be clear-cut at the age of 100. In this case, the forest owner would obtain from each stand a discounted stumpage value of \$ 16,000, with a probability of 0.04; of \$ 24,000, with a probability of 0.096, or \$ 70,000 with a probability of 0.864. The alternative of investing another 20 years in the 80 year old stand yields an expected value of \$ 63,424, with a standard deviation of \$ 16,629.

Cutting strategy 3

A third possibility is that the owner of the forest considers allowing the stands to grow for another 40 years, allowing the forest to reach an age of 120. In this case, the forest owner expects a discounted stumpage value of \$ 18,000 with a probability of 0.1; \$ 26,000, with a probability of 0.117; or \$ 76,000 with a probability of 0.783. For this investment, the expected value amounts to \$ 64,350, with a standard deviation of \$ 22,208.

The coefficient of correlation of the random income of mature spruce stands that reach an age of 100 years, but which could be maintained until an age of 120 years, is assumed to be 0.56. In order to simplify the problem, it can be assumed that the forest owner's total income from the first two strategies is invested in the forest enterprise and the rate of interest used is the internal rate of return.

Cutting Strategy	Rotation Age	Expected Stumpage Value, E	Standard deviation σ	Income Correlation		
				1	2	3
1	80	\$ 60,000	0	-	-	-
2	100	\$ 63,424	\$ 16,629		-	0.56
3	120	\$ 64,350	\$ 22,208			-

Table 4. Input data for an example in the determination of an optimal harvesting strategy.

The number of stands or the area that is to be harvested at 80, 100 and 120 years of age must now be determined in order to provide the highest possible risk utility for the forest owner. In this example, x_1 is the total area of 80 year old stands that are going to be harvested and the expected income per hectare of the cut is E_1. Similarly, x_2 and x_3 represents the area of 100 and 120 year old stands to be cut, and E_2 and E_3, their corresponding expected income per ha. The total area is designated by g. Once the variables x_2 and x_3 are determined, x_1 can be calculated through simple substraction ($x_1 = g - x_2 - x_3$).The expected value $E[V(x_2,x_3)]$ is calculated using eq. (10).

$$E[V(x_2,x_3)] = (g - x_2 - x_3) \cdot E_1 + x_2 \cdot E_2 + x_3 \cdot E_3 \quad (10)$$

To calculate the expected utility $W(x_2, x_3)$ using formula (9), the variance $Var[V(x_2, x_3)]$ is needed, in addition to the formula for the expected value $E[V(x_2, x_3)]$. The formula can be derived from the general variance formula (2):

$$Var[V(x_2,x_3)] = x_2^2 \cdot \sigma_2^2 + x_3^2 \cdot \sigma_3^2 + 2 \cdot x_2 \cdot x_3 \cdot \sigma_2 \cdot \sigma_3 \cdot \rho \quad (11)$$

To calculate the optimal areas, x_2 and x_3, for 100 and 120 year rotation stands, the function $W(x_2, x_3)$ must be expanded for these two variables:

$$W(x_2,x_3) = [(g - x_2 - x_3) \cdot E_1 + x_2 \cdot E_2 + x_3 \cdot E_3] - \frac{a}{2}(\cdot x_2^2 \cdot \sigma_2^2 + x_3^2 \cdot \sigma_3^2 + 2 \cdot x_2 \cdot x_3 \cdot \sigma_2 \cdot \sigma_3 \cdot \rho) \quad (12)$$

Differentiating $W(x_2, x_3)$ with respect to x_2 and x_3, yields the partial derivatives shown in eqs. (13) and (14) respectively, which are equations with the two unknown quantities x_2 and x_3.

$$\frac{dW(x_2,x_3)}{dx_2} = -E_1 + E_2 - a \cdot x_2 \cdot \sigma_2^2 - a \cdot x_3 \cdot \sigma_2 \cdot \sigma_3 \cdot \rho \quad (13)$$

$$\frac{dW(x_2,x_3)}{dx_3} = -E_1 + E_3 - a \cdot x_3 \cdot \sigma_3^2 - a \cdot x_2 \cdot \sigma_2 \cdot \sigma_3 \cdot \rho \quad (14)$$

Setting the derivatives equal to zero yields the necessary conditions for a maximum of the expected utility.

$$\frac{dW(x_2,x_3)}{dx_2} = -E_1 + E_2 - a \cdot x_2 \cdot \sigma_2^2 - a \cdot x_3 \cdot \sigma_2 \cdot \sigma_3 \cdot \rho = 0 \quad (15)$$

$$\frac{dW(x_2,x_3)}{dx_3} = -E_1 + E_3 - a \cdot x_3 \cdot \sigma_3^2 - a \cdot x_2 \cdot \sigma_2 \cdot \sigma_3 \cdot \rho = 0 \quad (16)$$

These conditions are also sufficient, because the second derivatives of $W(x_2,x_3)$ with respect to x_2 and x_3, are negative for $x_{2,3} \geq 0$, $a \geq 0$, $\sigma_{2,3} \geq 0$ and $\rho \geq 0$. By resolving either equation (15) or (16) and inserting one into the other, the formula for calculating x_{2opt} and x_{3opt} can be obtained. The solution of the problem is shown in the following steps.

$$-E_1 + E_3 - a \cdot x_3 \cdot \sigma_3^2 - a \cdot x_2 \cdot \sigma_2 \cdot \sigma_3 \cdot \rho = 0 \quad (17)$$

$$x_3 = \frac{-a \cdot x_2 \cdot \sigma_2 \cdot \sigma_3 \cdot \rho + (E_3 - E_1)}{a \cdot \sigma_3^2} \tag{18}$$

Now, x_3 can be substituted in eq. (15), which is subsequently made equal to zero:

$$(E_2 - E_1) - ax_2\sigma_2^2 - a\sigma_2\sigma_3\rho \cdot \frac{-ax_2\sigma_2\sigma_3\rho + (E_3 - E_1)}{a\sigma_3^2} = 0 \tag{19}$$

$$(E_2 - E_1) - ax_2\sigma_2^2 + \frac{ax_2\sigma_2^2\sigma_3\rho^2}{\sigma_3} - \frac{\sigma_2\rho(E_3 - E_1)}{\sigma_3} = 0 \tag{20}$$

$$ax_2\sigma_2^2\rho^2 - ax_2\sigma_2^2 = \frac{\sigma_2\rho(E_3 - E_1)}{\sigma_3} - (E_2 - E_1) \tag{21}$$

$$x_2\left(a\sigma_2^2\rho^2 - a\sigma_2^2\right) = -(E_2 - E_1) + (E_3 - E_1)\rho \cdot \frac{\sigma_2}{\sigma_3} \tag{22}$$

$$x_{2opt} = \frac{(E_3 - E_1) \cdot \rho \cdot \frac{\sigma_2}{\sigma_3} - (E_2 - E_1)}{a \cdot \sigma_2^2 \cdot \left(\rho^2 - 1\right)} \tag{23}$$

Similarly, x_{3opt} can be calculated as follows:

$$x_{3opt} = \frac{(E_2 - E_1) \cdot \rho \cdot \frac{\sigma_3}{\sigma_2} - (E_3 - E_1)}{a \cdot \sigma_3^2 \cdot \left(\rho^2 - 1\right)} \tag{24}$$

The optimal areas for harvesting, x_2 and x_3, can now be calculated since all input data required by eq. (23) and (24) are known. With this model, the forest owner is now able to choose optimal areas for each cutting strategy in order to achieve the highest expected utility with regard to his personal risk attitude. The input data of this example and the resulting optimal areas x_{2opt} and x_{3opt} are:

$E_1 = 60.000$

$E_2 = 63.424$ $\sigma_1 = 16.629$ $\sigma\rho = 0{,}56$

$E_3 = 64.350$ $\sigma_2 = 22.208$ $a = 1{,}666 \cdot 10^7$

$x_{2opt} = 51$

$x_{3opt} = 32$

$x_{1opt} = g - x_{2opt} - x_{3opt} = 17$

In order to achieve the maximum expected utility, an owner of a 100 hectare area of 80 year old forests, with risk tolerance a, should harvest 17 hectares immediately, at an age of 80; 51 hectares 20 years later, at an age of 100; and subsequently clear-cut the remaining 32 hectares after another 20 years, at an age of 120.

6.3 Example 2

In this example, the assumptions from example 1 will be altered and expanded in order to show the influence of different site qualities. Thus the stands do not possess the same expected values or dispersions of their stumpage values. Rather different risk classes are considered and consequently the same exploitation strategies lead to different expected values, dispersions and correlations. This assumption creates a more complex problem than example 1. In order to clearly illustrate the solution of the model, a very simple case is used as example 2.

The stands of a forest enterprise are to be assigned to two different risk classes,[4] 1 and 2. The total area of the mature stands, which are 80 years old, is 1,000 hectares. Of this area 300 hectares belong to risk class 1, which represents a risk for storm damage far below average; 700 hectares belong to risk class 2, which represents a storm damage rise above average for that area. The forest owner has to choose from two harvesting strategies to manage his 80 year old stands.

Cutting strategy 1

The first strategy is to allow stands to grow for another 20 years and clear-cut them at an age of 100. In this case, the forest owner expects a discounted stumpage value of $ 48,000 from low risk sites (risk class 1) and a stumpage value of $ 43,000 from the high risk sites (risk class 2). The standard deviation of the payments is $ 7,000 for risk class 1 and $ 8,000 for risk class 2.

Cutting strategy 2

The second strategy is to allow the stands to grow for another 40 years and to harvest them at an age of 120. The expected discounted stumpage values are $ 50,000 on the sites with risk class 1 and $ 47,000 on the sites with risk class 2; with standard deviations of $ 11,000 and $ 13,000 respectively.

The input data for example 2 is summarised in table 5, where:

B_1 = stands with a rotation age of 120 years on low risk sites.

B_2 = stands with a rotation age of 120 years on high risk sites.

B_3 = stands with rotation age of 100 years on low risk sites.

B_4 = stands with a rotation age of 100 years high risk sites.

[4] Sites with similar storm damage risk are grouped into risk classes, according to site characteristics and exposure of the area (König, 1996, p. 6).

	Age	Risk Class	Expected present stumpage value	Standard deviation σ	Income Correlation			
					B_1	B_2	B_3	B_4
B_1	120	low risk	\$ 50.000	\$ 11.000	-	0,70	0,50	0,35
B_2	120	high risk	\$ 47.000	\$ 13.000		-	0,40	0,65
B_3	100	low risk	\$ 48.000	\$ 7.000			-	0,60
B_4	100	high risk	\$ 43.000	\$ 8.000				-

Table 5. Input data of example 2, the determination of an optimal harvesting strategy for 80 year old mature stands, on an area of 1,000 hectares that has been classified into two risk classes.

The stumpage value of one hectare of an 80 year old stand is \$ 40,000. The value b, for all stands, is therefore \$ 40,000 · 1.000 = \$ 4 · 10^7. In example 2, the forest owner is presumably less risk tolerant than in example 1, with a = 1,5 / b = 3,75 · 10^{-8}. The expected utility $W(x_1,...,x_4)$ can be calculated using this data and formula (9). The variables x_1, x_2, x_3 and x_4 are defined as follows:

x_1 = Area of stands with a rotation age of 120 years on low risk sites (B_1).

x_2 = Area of stands with a rotation age of 120 years on high risk sites (B_2).

x_3 = Area of stands with a rotation age of 100 years on low risk sites (B_3).

x_4 = Area of stands with a rotation age of 100 years on high risk sites (B_4).

For this example, the formula for the expected utility is:

$$W(x_1, x_2, x_3, x_4) = E_1 \cdot x_1 + E_2 \cdot x_2 + E_3 \cdot x_3 + E_4 \cdot x_4 - \frac{a}{2} \cdot [\mathrm{I}+\mathrm{II}+\mathrm{III}] \quad (25)$$

$$(I) = x_1^2\sigma_1^2 + x_2^2\sigma_2^2 + x_3^2\sigma_3^2 + x_4^2\sigma_4^2$$

$$(II) = 2x_1x_2\sigma_1\sigma_2\rho_{1/2} + 2x_1x_3\sigma_1\sigma_3\rho_{1/3} + 2x_1x_4\sigma_1\sigma_4\rho_{1/4}$$

$$(III) = 2x_2x_3\sigma_2\sigma_3\rho_{2/3} + 2x_2x_4\sigma_2\sigma_4\rho_{2/4} + 2x_3x_4\sigma_3\sigma_4\rho_{3/4}$$

In order to estimate the values of x_1, x_2, x_3 and x_4, which yield the highest expected utility, the partial derivatives of $W(x_1,..., x_4)$ with respect to x_1, x_2, x_3 and x_4 are calculated. The four derivatives are:

$$\frac{dW(x)}{d(x_1)} = E_1 - \frac{a}{2}(2\sigma_1^2 x_1 + 2x_2\sigma_1\sigma_2\rho_{1/2} + 2x_3\sigma_1\sigma_3\rho_{1/3} + 2x_4\sigma_1\sigma_4\rho_{1/4}) \quad (26)$$

$$\frac{dW(x)}{d(x_2)} = E_2 - \frac{a}{2}(2\sigma_2^2 x_2 + 2x_1\sigma_1\sigma_2\rho_{1/2} + 2x_3\sigma_2\sigma_3\rho_{2/3} + 2x_4\sigma_2\sigma_4\rho_{2/4}) \quad (27)$$

$$\frac{dW(x)}{d(x_3)} = E_3 - \frac{a}{2}(2\sigma_3^2 x_3 + 2x_1\sigma_1\sigma_3\rho_{1/3} + 2x_2\sigma_2\sigma_3\rho_{2/3} + 2x_4\sigma_3\sigma_4\rho_{3/4}) \quad (28)$$

$$\frac{dW(x)}{d(x_4)} = E_4 - \frac{a}{2}(2\sigma_4^2 x_4 + 2x_1\sigma_1\sigma_4\rho_{1/4} + 2x_2\sigma_2\sigma_4\rho_{2/4} + 2x_3\sigma_3\sigma_4\rho_{3/4}) \quad (29)$$

If all derivatives are made equal zero, the following linear equation system can be solved:

$$\frac{E_1}{\sigma_1 a} = \sigma_1 x_1 + \sigma_2\rho_{1/2}x_2 + \sigma_3\rho_{1/3}x_3 + \sigma_4\rho_{1/4}x_4 \quad (30)$$

$$\frac{E_2}{\sigma_2 a} = \sigma_1\rho_{1/2}x_1 + \sigma_2 x_2 + \sigma_3\rho_{2/3}x_3 + \sigma_4\rho_{2/4}x_4 \quad (31)$$

$$\frac{E_3}{\sigma_3 a} = \sigma_1\rho_{1/3}x_1 + \sigma_2\rho_{2/3}x_2 + \sigma_3 x_3 + \sigma_4\rho_{3/4}x_4 \quad (32)$$

$$\frac{E_4}{\sigma_4 a} = \sigma_1\rho_{1/4}x_1 + \sigma_2\rho_{2/4}x_2 + \sigma_3\rho_{3/4}x_3 + \sigma_4 x_4 \quad (33)$$

This linear set of equations can be solved with the help of the matrix calculus, as follows:

$$\boldsymbol{A} = \begin{pmatrix} \sigma_1 & \sigma_2\rho_{1/2} & \sigma_3\rho_{1/3} & \sigma_4\rho_{1/4} \\ \sigma_1\rho_{1/2} & \sigma_2 & \sigma_3\rho_{2/3} & \sigma_4\rho_{2/4} \\ \sigma_1\rho_{1/3} & \sigma_2\rho_{2/3} & \sigma_3 & \sigma_4\rho_{3/4} \\ \sigma_1\rho_{1/4} & \sigma_2\rho_{2/4} & \sigma_3\rho_{3/4} & \sigma_4 \end{pmatrix}; \boldsymbol{x} = \begin{pmatrix} x_1 \\ x_2 \\ x_3 \\ x_4 \end{pmatrix}; \boldsymbol{c} = \begin{pmatrix} E_1 / \sigma_1 a \\ E_2 / \sigma_2 a \\ E_3 / \sigma_3 a \\ E_4 / \sigma_4 a \end{pmatrix};$$

Using an algebraic matrix, the equation system (30) to (33) can be represented as follows:

$$\boldsymbol{A} \cdot \boldsymbol{x} = \boldsymbol{c} \quad (34)$$

The solution to vector **x** is obtained by pre-multiplying[2] the equation with the inverse $\mathbf{A}^{-1}$, resulting in:

[2] For details of matrix algebra, see Bortz (1993, p. 635 ff.)

$$A^{-1}\sigma \cdot A \cdot x = A^{-1}\sigma \cdot c \qquad (35)$$

$$x = A^{-1}\sigma \cdot c \qquad (36)$$

A solution of this equation system may generally result in both positive and negative values for x_1, x_2, x_3 and x_4. Since the forest area is exactly 1,000 hectares, the sum of x_1, x_2, x_3 and x_4 must amount to 1,000 and in addition negative values are not feasible. Furthermore, stands from risk class 1 must occupy an area of 300 hectares, while stands from risk class 2 must cover an area of 700 hectares. These additional restrictions can be expressed in the following two equations:

$$x_1 + x_3 = 300 \qquad (37)$$

$$x_2 + x_4 = 700 \qquad (38)$$

This last restriction marks the difference between a forest manager and an asset manager who invests in shares. The latter is able to acquire any amount of each share, until he reaches his financial restriction, while the harvesting strategy of the forest owner is restricted by the distribution of stands into risk classes.

6.3.1 Non-linear optimisation

In addition to the expected utility objective function, *W(x)* (25), the two restrictions (37) and (38) must also be considered in the solution of example 2. Suitable methods for optimising such objective functions with constraints, arose in the 1930s (Zimmermann, 1992,p. 5) and later developed into the scientific discipline called Operation Research. The following solution to example 2 was originally presented by Zimmermann (1992, p. 152 ff).

In order to facilitate further work with objective function and constraints, all equations will be written in matrix form.

For the expected utility, $W(x) = W(x_1, x_2, x_3, x_4)$, it follows that:

$$W(x)=(E_1,E_2,E_3,E_4)\cdot\begin{pmatrix}x_1\\x_2\\x_3\\x_4\end{pmatrix}-\frac{a}{2}\cdot(x_1,x_2,x_3,x_4)$$

$$\cdot\begin{pmatrix}\sigma_1^2 & \sigma_2\sigma_1\rho_{1/2} & \sigma_3\sigma_1\rho_{1/3} & \sigma_4\sigma_1\rho_{1/4}\\ \sigma_1\sigma_2\rho_{1/2} & \sigma_2^2 & \sigma_3\sigma_2\rho_{2/3} & \sigma_4\sigma_2\rho_{2/4}\\ \sigma_1\sigma_3\rho_{1/3} & \sigma_2\sigma_3\rho_{2/3} & \sigma_3^2 & \sigma_4\sigma_3\rho_{3/4}\\ \sigma_1\sigma_4\rho_{1/4} & \sigma_2\sigma_4\rho_{2/4} & \sigma_3\sigma_4\rho_{3/4} & \sigma_4^2\end{pmatrix}\cdot\begin{pmatrix}x_1\\x_2\\x_3\\x_4\end{pmatrix}$$

In the short form of the matrix notation, this equation can be expressed:

$$W(x)=\boldsymbol{E}^T\boldsymbol{x}-\frac{a}{2}\,x^TQ\,x \qquad (39)$$

The two constraints that exactly 300 hectares of forest are on low risk sites and 700 hectares of forest are on high risk sites are:

$$\begin{matrix}x_1+x_3=300\ Hectare=d\\ x_2+x_4=700\ Hectare=e\end{matrix}\quad\Leftrightarrow\quad\begin{pmatrix}1&0&1&0\\0&1&0&1\end{pmatrix}\cdot\begin{pmatrix}x_1\\x_2\\x_3\\x_4\end{pmatrix}=\begin{pmatrix}d\\e\end{pmatrix}$$

This can be simplified and written as

$$\boldsymbol{A}\cdot\boldsymbol{x}=\boldsymbol{b} \qquad (40)$$

In order to check its validity, (40) can be expanded:

$$1\,x_1 \quad +0\,x_2 \quad +1\,x_3 \quad +0\,x_4 \quad =x_1+x_3 \quad =d$$

$$0\,x_1 \quad +1\,x_2 \quad +0\,x_3 \quad +1\,x_4 \quad =x_2+x_4 \quad =e$$

Since stand areas cannot take negative values, the non-negative condition $\boldsymbol{x}\geq 0$ also is necessary.

The objective function $W(x)$ (39) is a quadratic function, but the distribution constraints (40) are modelled as a linear equation system. It is therefore necessary to use a non-linear optimisation selection procedure, which finds a solution to the model (39) under distribution (40) and non-negativity ($x\geq 0$) restrictions. Such a procedure is called the Wolfe algorithm (Zimmermann, 1992, p. 159). This algorithm can only minimise the following type of objective function.

Min. $$W(x)=c^T\cdot\boldsymbol{x}+\frac{1}{2}\cdot\boldsymbol{x}^T\cdot Q\cdot\boldsymbol{x} \qquad (41)$$

Additionally the constraints must be expressed in the following form:

$$\boldsymbol{A} \cdot \boldsymbol{x} \leq \boldsymbol{b} \tag{42}$$

and it must follow that:

$$\boldsymbol{x} \geq \boldsymbol{0} \tag{43}$$

The matrix $\boldsymbol{Q}$ thereby must be

$\boldsymbol{Q}$ = symmetrical and positively semi-definite (44)

Firstly, equation (39) should be transformed into equation (41). Since the objective is to maximize (39) and the Wolfe algorithm performs only minimisation, equation (39) must be multiplied by -1 (Zimmermann, 1992, p. 81). The following relationship applies:

$$\text{maximize} \quad W(x) \quad = \quad \text{minimize} \quad -W(x) \quad \Rightarrow$$

$$\text{maximize} \quad \boldsymbol{E}^T x - \frac{a}{2} x^T \boldsymbol{Q} x \quad = \quad \text{minimize} \quad -\boldsymbol{E}^T x + \frac{a}{2} x^T \boldsymbol{Q} x$$

If $\boldsymbol{E}^{*T} = -\boldsymbol{E}^T$ and $\boldsymbol{Q}^* = a\,\boldsymbol{Q}$, the following applies to equation (39)[3]

$$\text{maximize} \ \boldsymbol{E}^T x - \frac{a}{2} x^T \boldsymbol{Q} x \quad = \quad \text{minimize} \quad \boldsymbol{E}^{*T} x + \frac{1}{2} x^T \boldsymbol{Q}^* x$$

Thus, (39) fulfils the prerequisite of the algorithm (41) and conditions (42) and (43) are likewise fulfilled. Now it must be proven that $\boldsymbol{Q}^*$ is symmetrical and positively semi-definite. By examining the matrix:

$$\boldsymbol{Q} = \begin{pmatrix} \sigma_1^2 & \sigma_2\sigma_1\rho_{1/2} & \sigma_3\sigma_1\rho_{1/3} & \sigma_4\sigma_1\rho_{1/4} \\ \sigma_1\sigma_2\rho_{1/2} & \sigma_2^2 & \sigma_3\sigma_2\rho_{2/3} & \sigma_4\sigma_2\rho_{2/4} \\ \sigma_1\sigma_3\rho_{1/3} & \sigma_2\sigma_3\rho_{2/3} & \sigma_3^2 & \sigma_4\sigma_3\rho_{3/4} \\ \sigma_1\sigma_4\rho_{1/4} & \sigma_2\sigma_4\rho_{2/4} & \sigma_3\sigma_4\rho_{3/4} & \sigma_4^2 \end{pmatrix}$$

it is clear that the terms on either side of the diagonal are identical, making this a symmetrical matrix. Since in $\boldsymbol{Q}^*$, $\boldsymbol{Q}$ is multiplied by the constant a, $\boldsymbol{Q}^*$ is also symmetrical (Bortz, 1993, p. 636). Because risk aversion $a \geq 0$, all standard deviations $\sigma_i \geq 0$, and all coefficients of

[3] From the vector of the expected values $\boldsymbol{E}^T$ a new vector emerges by the transformation to $\boldsymbol{E}^{*T}$, in which the expected values are assigned negative signs. All coefficients of the matrix $\boldsymbol{Q}$ are multiplied by a, resulting in the matrix $\boldsymbol{Q}^*$. Model (41) is obtained with $\boldsymbol{E}^{*T}$ and $\boldsymbol{Q}^*$.

correlation $\rho_{ij} \geq 0$, it follows that Q^* must be positively semi-definite (Zimmermann, 1992, p. 157). Thus, the objective function is convex[4] and its minimum is the global minimum.

In the next step, the quadratic objective function must be converted into a linear function, for which the Lagrange-function is used. If the objective is to minimise $W(x)$, under the restriction $g_i(x) \leq \mathbf{0}$ and $x \geq 0$, then the function $L(x,\boldsymbol{u}) = W(x) + \boldsymbol{u}^T g_i(x)$ is the appropriate Lagrange-function. Kuhn and Tucker formulated the necessary and sufficient conditions for the vector $(\boldsymbol{x}_0, \boldsymbol{u}_0)^T$ to be a saddle point of the Lagrange-function. If $(\boldsymbol{x}_0, \boldsymbol{u}_0)^T$ for $x_0 \geq 0$ and $u_0 \geq 0$ is a saddle point of L, then x_0 is also an optimal solution for the model "minimise $W(x)$ with $g_i(x) \leq 0$ and $x \geq \mathbf{0}$" (Zimmermann, 1992, p. 154).

For equations (41) to (44), the Lagrange-function and Kuhn-Tucker conditions (Zimmermann, 1992, p. 155 f.) are the following. $\boldsymbol{L_x}$ designates the first partial derivative of $L(\boldsymbol{x}, \boldsymbol{u})$ with respect to $\boldsymbol{x}$ and $\boldsymbol{L_u}$ designates the first derivative of $L(x, \boldsymbol{u})$ with respect to the Lagrange multiplier $\boldsymbol{u}$.

Lagrange-Function $$L(x,u) = c^T x + \frac{1}{2} x^T Q x + u^T (A x - b) \quad (45)$$

Kuhn-Tucker-conditions

$$\boldsymbol{L_x} = \boldsymbol{c} + \boldsymbol{Q x} + \boldsymbol{A}^T \boldsymbol{u} \geq \mathbf{0} \quad (45.1)$$

$$\boldsymbol{L_u} = \boldsymbol{A x} - \boldsymbol{b} \leq \mathbf{0} \quad (45.2)$$

$$\boldsymbol{L_x x_0} = 0 \quad (45.3)$$

$$\boldsymbol{L_u u_0} = 0 \quad (45.4)$$

$$\boldsymbol{u} \geq \mathbf{0}, \boldsymbol{x} \geq \mathbf{0} \quad (45.5)$$

By formulating the Kuhn-Tucker conditions, a linear equation system was created from the quadratic Lagrange-function. Subsequently, it must be tested whether the Kuhn-Tucker conditions can be fulfilled. In this case the vector $\boldsymbol{x_0}$ contains the optimal solution. The Kuhn-Tucker conditions are extended by the slack variables[5] $\boldsymbol{s_1}$ and $\boldsymbol{s_2}$, and under the consideration of the so-called complementary conditions (45.3) and (45.4), the following equation system is created:

$$\boldsymbol{c} + \boldsymbol{Q x} + \boldsymbol{A}^T \boldsymbol{u} - \boldsymbol{s_1} = 0 \quad (46.1)$$

$$-\boldsymbol{b} + \boldsymbol{A x} \qquad + \boldsymbol{s_2} = 0 \quad (46.2)$$

---

[4] A set $M \in \mathrm{IR}^n$ is called convex, if with each of the two points $x_1, x_2 \in M$, every convex linear combination of x_1 and x_2 belongs to M (ZIMMERMANN, 1992, p. 153). In other words, a convex quantity M is characterised by the fact that each direct connection between two points lies completely within M.

[5] The so-called slack variables are introduced so that an equation system can be created from the inequality system. For instance, in optimisations they can be regarded as underutilization of capacities (BLOECH, 1974, p. 48).

Equations (46.1) and (46.2) can now be solved with the help of a Simplex tableau,[6] as long as a feasible optimal basic solution exists. The next step is to determine the Kuhn-Tucker conditions. For simpler presentation Q^* will be rewritten:

$$Q^* = \begin{pmatrix} a\sigma_1^2 & a\sigma_2\sigma_1\rho_{1/2} & a\sigma_3\sigma_1\rho_{1/3} & a\sigma_4\sigma_1\rho_{1/4} \\ a\sigma_1\sigma_2\rho_{1/2} & a\sigma_2^2 & a\sigma_3\sigma_2\rho_{2/3} & a\sigma_4\sigma_2\rho_{2/4} \\ a\sigma_1\sigma_3\rho_{1/3} & a\sigma_2\sigma_3\rho_{2/3} & a\sigma_3^2 & a\sigma_4\sigma_3\rho_{3/4} \\ a\sigma_1\sigma_4\rho_{1/4} & a\sigma_2\sigma_4\rho_{2/4} & a\sigma_3\sigma_4\rho_{3/4} & a\sigma_4^2 \end{pmatrix} = \begin{pmatrix} q_{11} & q_{12} & q_{13} & q_{14} \\ q_{21} & q_{22} & q_{23} & q_{24} \\ q_{31} & q_{32} & q_{33} & q_{34} \\ q_{41} & q_{42} & q_{43} & q_{44} \end{pmatrix}$$

The remaining coefficients can be written as vectors or as matrices:

$$A = \begin{pmatrix} 1 & 0 & 1 & 0 \\ 0 & 1 & 0 & 1 \end{pmatrix}; \qquad A^T = \begin{pmatrix} 1 & 0 \\ 0 & 1 \\ 1 & 0 \\ 0 & 1 \end{pmatrix};$$

$$u = \begin{pmatrix} u_1 \\ u_2 \end{pmatrix}; \quad A^T u = \begin{pmatrix} u_1 \\ u_2 \\ u_3 \\ u_4 \end{pmatrix}; \quad b = \begin{pmatrix} d \\ e \end{pmatrix}; \quad E^{*T} = (-E_1, -E_2, -E_3, -E_4)$$

Thus, the Kuhn-Tucker conditions are:

$$L_x = \begin{Bmatrix} -E_1 & +q_{11}x_1 & +q_{12}x_2 & +q_{13}x_3 & +q_{14}x_4 & +u_1 & \geq 0 \\ -E_2 & +q_{21}x_1 & +q_{22}x_2 & +q_{23}x_3 & +q_{24}x_4 & +u_2 & \geq 0 \\ -E_3 & +q_{31}x_1 & +q_{32}x_2 & +q_{33}x_3 & +q_{34}x_4 & +u_3 & \geq 0 \\ -E_4 & +q_{41}x_1 & +q_{42}x_2 & +q_{43}x_3 & +q_{44}x_4 & +u_4 & \geq 0 \end{Bmatrix}$$

$$L_u = \begin{Bmatrix} x_1 & & +x_3 & & -d & 0 \\ & x_2 & & +x_4 & -e & 0 \end{Bmatrix}$$

$$\boldsymbol{L_x}\ \boldsymbol{x_0} = 0$$

$$\boldsymbol{L_u}\ \boldsymbol{u_0} = 0$$

$\boldsymbol{L_x}$ and $\boldsymbol{L_u}$ can be rearranged and the slack variables s_{11} to s_{14}, s_{21} and s_{22} added. Thus, the following six equations are obtained, which will be the starting base for the first simplex tableau:

[6] Detailed explanations of the Simplex algorithm can be found in ZIMMERMANN (1992, p. 67 ff.) and BLOECH (1974, p. 47 ff.).

$$
\begin{array}{lllllllllllll}
q_{11}x_1 & + q_{12}x_2 & + q_{13}x_3 & + q_{14}x_4 & + u_1 & - s_{11} & & & & & & = E_1 \\
q_{21}x_1 & + q_{22}x_2 & + q_{23}x_3 & + q_{24}x_4 & + u_2 & & - s_{12} & & & & & = E_2 \\
q_{31}x_1 & + q_{32}x_2 & + q_{33}x_3 & + q_{34}x_4 & + u_1 & & & - s_{13} & & & & = E_3 \\
q_{41}x_1 & + q_{42}x_2 & + q_{43}x_3 & + q_{44}x_4 & + u_2 & & & & - s_{14} & & & = E_4 \\
x_1 & & + x_3 & & & & & & & + s_{21} & & = d \\
& x_2 & & + x_4 & & & & & & & + s_{22} & = e
\end{array}
$$

After inserting the values for q_{ij}, E_i, d and e, this equation system can be written as:

$$
\begin{array}{lllllllllllll}
4{,}538\,x_1 & + 3{,}754\,x_2 & + 1{,}444\,x_3 & + 1{,}155\,x_4 & + u_1 & - s_{11} & & & & & & = 50.000 \\
3{,}754\,x_1 & + 6{,}338\,x_2 & + 1{,}365\,x_3 & + 2{,}535\,x_4 & + u_2 & & - s_{12} & & & & & = 47.000 \\
1{,}444\,x_1 & + 1{,}365\,x_2 & + 1{,}838\,x_3 & + 1{,}26\,x_4 & + u_1 & & & - s_{13} & & & & = 48.000 \\
1{,}155\,x_1 & + 2{,}535\,x_2 & + 1{,}26\,x_3 & + 2{,}4\,x_4 & + u_2 & & & & - s_{14} & & & = 43.000 \\
x_1 & & + x_3 & & & & & & & + s_{21} & & = 300 \\
& x_2 & & + x_4 & & & & & & & + s_{22} & = 700
\end{array}
$$

The solution of this system of equations is shown in the appendix.

The simplex algorithm can be used only, if a linear objective function and a set of linear constraints are given. In business management, an optimal solution, frequently the maximum gross profit (Bloech, 1974, p. 49), is always searched for in one of the corners of the convex set of restrictions. This is because the objective function possesses a maximum in at least one corner of the convex set of constraints (polyhedron of the restrictions) (Bloech, 1974, p. 49).

Graphically speaking, the simplex algorithm moves from the coordinate origin through the corners of the polyhedron. During this process the corner that yields the maximum objective function value increase, is chosen. This process guarantees an consistent search for the optimal solution. Variables which are not in the basis of the simplex tableau are predetermined zero in every corner. Details regarding base variables and non-base variables are given by Bloech (1974, p. 55 ff).

In our example, the Kuhn-Tucker conditions (stated above) form the restriction polyhedron. Since the simplex algorithm starts in the coordinate origin (tableau 1), the variables x_1 to x_4, u_1 and u_2 are set at zero and the variable s_{11} to s_{14}, s_{21} and s_{22} are the base variables. The base variables do not form an identity matrix[7] because s_{11} to s_{14} have minus signs. For this reason, the set of equations should be extended around the so-called auxiliary variables h_{11} to h_{14} (Zimmermann, 1992, p. 76 f.). In view of the optimisation, $-M$ is a very unfavourable value for the auxiliary variables. M is a very large number in comparison with c_i (Zimmermann, 1992, p. 77). The coefficients of the objective function results from the introduction of the artificial variables and their value - M, as can be seen in tableau 1 (see appendix 1).

A continual maximal improvement of the base solution from tableau to tableau is guaranteed by the so-called acceptance rule (Zimmermann, 1992, p. 72). The non-base variable, whose ΔZ_j (last row of each tableau) possesses the smallest negative value, causes the highest increase in the objective function's value. ΔZ_j can be considered as the opportunity cost or shadow price (Bloech, 1974, p. 69). The choice of the variable with the highest shadow price (smallest negative ΔZ_j) as a part of the base in the next step, givesthe highest increase in the objective function's value, as long as the variable is predetermined as zero. Since the shadow prices have negative signs at the beginning of the optimisation process and after each change of base, the improvements of ΔZ_j are added to the values of the non-base variables and the optimal base solution is achieved when all $\Delta Z_j \geq 0$. The column of the variable with the smallest negative ΔZ_j is called pivot column.

The variable that must be removed from the base and be replaced by a new variable is determined using the lowest value of Θ, the last column of each tableau, which must be greater than zero. This rule, called the elimination rule, ensures that the solution is valid (Zimmermann, 1992, p. 68) and thus prevents negative values for the base variables. The row with the lowest value for $\Theta > 0$ is called pivot row.

The element of the simplex tableau that belongs to both the pivot column and the pivot row is called pivot element. The pivot element is used for the formation of the inverse matrix and thus for the solution of the linear equation system. Details for the calculation of inverses can be found in Bloech (1974, p. 52 ff). If the tableaus are pivoted, then the identity matrix will be created based on the matrix of the coefficients of the prevailing base variables. Thus, the inverse of the base variable coefficient matrix is obtained, with which the old solution vector b

[7] A matrix whose main diagonal is made up of ones with remaining elements equal to 0, is called identity (unit) matrix $\boldsymbol{I}$ (Bortz, 1993, p. 637).

is pre-multiplied. The result of this calculation is the new base solution.The pivoting is performed according to the following rule:

The elements of the pivot rows are divided by the pivot element. Then, from the elements of the other rows, a multiple of elements of the pivot rows should be subtracted, so that the element in the pivot column becomes zero.

Pivoting is demonstrated for Tableau 1 in the appendix.

6.3.2 Results

Tableau 7 (see appendix) presents the optimal solution for example 2, which will be briefly outlined again.

A forest owner possesses 1,000 hectares of 80 year old mature stands, of which 300 hectares are on low risk sites and 700 hectares on high risk sites. The stands can be harvested either at an age of 100 or 120. Depending on the rotation age and risk of storm damage, a stumpage value with a known dispersion can be expected for each stand. Additionally, the correlations of the stumpage values of the different stands are known values. With the method presented, the optimal harvesting strategy given by the maximum expected utility $W(x)$, can be calculated for the forest owner with a given risk tolerance. The required variables x_1, x_2, x_3 and x_4 are defined as follows:

$x_1 =$ Area of stands with rotation age of 120 years on low risk sites.

$x_2 =$ Area of stands with rotation age of 120 years on high risk sites

$x_3 =$ Area of stands with rotation age of 100 years on low risk sites.

$x_4 =$ Area of stands with rotation age of 100 years on high risk sites.

The results of the non-linear optimisation process are:

$x_1 = 128$ $\quad x_2 = 700$ $\quad x_3 = 172$ $\quad x_4 = 0$

The forest owner achieves the maximum expected utility, if all stands on high risk sites are harvested at an age of 120 years. On low risk sites 128 hectares and 172 hectares are cut at an age of 120 and 100 respectively. The maximum expected utility as a result of this exploitation strategy is \$ 45,406,493. The expected value for the optimal combination of x_1, x_2, x_3 and x_4, which does not consider risk attitude, is \$ 47,555,771.

The reader might be surprised that the optimal cutting strategy in this example suggests the harvest of all stands on high risk sites at a later date, while a portion of the stands on low risk sites should be harvested earlier. Intuitively, the opposite strategy would be considered best, whereupon stands on higher risk sites should be harvested earlier in order to minimise risk. The calculated result can be explained with the input data from table 5. In the 20 year growth period between 100 and 120 the expected present stumpage value increases by about $ 2,000 for stands on low risk sites, while the value of the stands on high risk sites increases about $ 4,000. The standard deviations of the expected values increase for stands on high risk sites, by $ 1,000 more than for stands on low risk sites. However, the variance enters the expected utility objective function, reduced by a/2. This increase affects the expected utility less than the increase of the expected value. The calculated results are only valid with the presented input data and should not be generalised.

7 Conclusions, discussion and outlook

Financial theory has developed many suitable risk incorporation models, which may be especially helpful in supporting silvicultural decisions in forest enterprises, when considering risks in forest management. However, until now these models have not been used in forest management practice in Germany, although the preconditions for the use of advanced planning techniques do exist. For example, larger private and community owned forest enterprises are legally forced to provide medium-term management plans and the state owned forest enterprises have specialised departments for forest planning. It is thought that the use of advanced planning models violates the traditional forest planning procedure practised in Germany. In traditional forest planning, firstly plans are made independently for each stand and then the overall plan is made by combining the individual plans. This plan is often subject to further minor revisions. The individual risks to which each stand is subject to is undoubtedly considered with this procedure. However, this is done intuitively by the forester who is preparing the inventory and plan for the stand. Therefore, how exactly the risks are considered in this procedure is not clear. Certainly, risk consideration depends strongly on the respective planner's risk tolerance and hence is distributed over the forest area as the risk attitudes of the planners. Formal and conscious risk management has not emerged in Germany and practically no forest manager has quantitative knowledge of the risks of his investments. But it is supposed that substantial unsatisfied demand for improved planning tools exist in this area. There is certainly a readiness for the application of modern planning models, but there is also a substantial lack of quantitative data on risk (v. Gadow, 2000). While attempts to increase for the quantitative understanding of risk begun much earlier in other countries, e.g. the United Kingdom, in Germany, interest was directed to the quantification of the forest management risks only after the storm damage that occurred in 1990. Although some research projects have been carried out in the last decade, it cannot be claimed that much emphasis was placed on these efforts. After the considerable storm damage that occurred yet again in 1999, it is expected that empirical research in this area will be continued.

In the past years, the behaviour of the public forest sector regarding risks in forest management and risk control, has been contradictory. Risk has often been used to justify silvicultural operations, especially concerning the transformation from coniferous forest stands to broadleaved (hardwood) forest stands. On the other hand, decisions have been made that probably increase risks considerably, in particular, the extension of rotation periods. A significant reduction of storm damage risk could be obtained by cutting high risk spruce stands relatively early, while for other stands on low risk sites longer rotations could be choosen. However so far resolute actions have not been taken. It may be presumed that risk reduction is only an ostensible argument, with which other reasons for the promotion of hardwood are justified. In this case, conscious risk management is not achieved.

Another noteworthy aspect is the uncertainty of future timber prices. The risk caused by market development is a good reason for forest enterprises to invest in a combination of tree species. However, it seems that such decisions are generally made intuitively, since there are practically no studies on the correlation between timber prices and the effects of diversification.

K. v. Gadow (2000) stresses that applications of risk analysis are surprisingly rare in forest planning. Hence, the question is, which obstacles are opposed to the use of decision models that consider risk, like those presented in this paper, and whether and how these obstacles can be overcome. Principally, consideration of risk requires the use of suitable decision support models from the field of Operation Research. The obstacles are therefore comparable to those restricting the use of OR procedures for decision support in many other fields. The application of such procedures is only possible if either the decision problem is of such importance that the decision maker is willing to invest time and resources into constructing a sophisticated model, or the decision problem must be solved so frequently that programming a solution with user-friendly interface is worth providing. The programming should permit the use of the model without any knowledge on the algorithms employed by the program. For example, many programs for optimisation of planning routes are used and the users are given no information on the calculation procedure leading to a suggested route. In the case of forest enterprise planning, decisions are certainly relatively frequent and suitable for standardising, therefore a routine solution could be prepared. Such a computer program could be used by the respective decision makers without requiring detailed knowledge of the actual optimisation process. However, in contrast to the optimisation routing mentioned above as an example, the acceptance of planning results could be limited. Both the problem and the solution method are very complex and can be difficult to understand. Moreover, judging the results with common sense is also fairly complicated. However, the situation is not hopeless and it is most likely that decision support models very much like those presented here, may be used in the future. This is because the technical prerequisites will be improved by better availability of high performance computers. Also, future generation decision makers will be more accustomed to the use of computer models.

References

Bloech, J., 1974: Lineare Optimierung für Wirtschaftswissenschaftler. Studienbücher BWL/ VWL Band 2. Opladen: Westdeutscher Verlag.

Blohm, H.; Lüder, K., 1991: Investition, 7. Auflage, Vahlen, München.

Bortz, J., 1993: Statistik für Sozialwissenschaftler. Berlin, Heidelberg: Springer Verlag.

Dieter, M., 1997: Berücksichtigung von Risiko bei forstbetrieblichen Entscheidungen. Schriften zur Forstökonomie, Band 16, Sauerländer's Verlag, Frankfurt am Main.

Dieter, M., 1999: Betriebswirtschaftliche Untersuchungen zum Voranbau in Fichtenaltbeständen. Forstwissenschaftliches Centralblatt (118. Jg.), S. 145-155.

Gadow, K.v., 2000: Evaluating Risk in Forest Planning Models. Silva Fennica 34 (2), (in print)

König, A., 1995: Sturmgefährdung von Beständen im Altersklassenwald. Ein Erklärungs-und Prognosemodell. Frankfurt/Main: J.D. Sauerländer's.

König, A., 1996: Unveröffentlichter Abschlußbericht des Projektes: Abgrenzung von Sturmschadensrisikoklassen und Entwicklung von risikoorientierten Endnutzungsstrategien für den bayerischen Staatswald. Bayerische Landesanstalt für Wald und Forstwirtschaft, Freising.

König, A., Mößmer, R., Bäumler, A., 1995: Waldbauliche Dokumentation der flächigen Sturmschäden des Frühjahres 1990 in Bayern und meteorologische Situation zur Schadenszeit. Berichte aus der Landesanstalt für Wald und Forstwirtschaft, Freising, Band 2.

Kruschwitz, L., 1995: Investitionsrechnung. de Gruyter, Berlin.

Möhring, B., 1986: Dynamische Betriebsklassensimulation. Berichte des Forschungszentrums Waldökosysteme-Waldsterben, Band 20, Göttingen.

Richter, J., 1996: Sturmschäden in Fichtenbeständen. In: Allgemeine Forst- und Jagdzeitung (167. Jg.), Heft 12, S. 234-238.

Richter, J., 1998: Überschätzter HD-Wert? In: AFZ/Der Wald, Heft 15, S. 791-792.

Spremann, K., 1991: Investition und Finanzierung. München, Wien: Oldenbourg Verlag.

Waldherr, M., 1997: Das Schadholzrisiko im Staatswald der Forstdirektion Niederbayern-Oberpfalz während der vergangenen 40 Jahre. In: Forst und Holz (52. Jg.), Nr. 2, S. 27-30. Hannover: M. & H. Schaper.

Zimmermann, H. J., 1992: Methoden und Modelle des Operations Research. Braunschweig, Wiesbaden: Friedrich Vieweg & Sohn.

Appendix

Simplex-tableaus for the solution to example 2

Tableau 1

The first tableau contains the auxiliary variables h_{11} to h_{14} and the slack variables s_{21} and s_{22} in the basis. The nonbasis variable with the minimum value of ΔZ_j is x_2. Thus the pivot column is determined. By dividing b by the coefficients of x_2, Θ results. The variable s_{22} has the least positive value of Θ.

	c_i	x_1	x_2	x_3	x_4	u_1	u_2	s_{11}	s_{12}	s_{13}	s_{14}	s_{21}	s_{22}	h_{11}	h_{12}	h_{13}	h_{14}	b	Θ	
h_{11}	- M	4,538	3,754	1,444	1,155	1	0	-1	0	0	0	0	0	1	0	0	0	50.000	13.320	
h_{12}	- M	3,754	6,338	1,365	2,535	0	1	0	-1	0	0	0	0	0	1	0	0	47.000	7.416	
h_{13}	- M	1,444	1,365	1,838	1,26	1	0	0	0	-1	0	0	0	0	0	1	0	48.000	35.165	
h_{14}	- M	1.155	2,535	1,26	2,4	0	1	0	0	0	-1	0	0	0	0	0	1	43.000	16.963	
s_{21}	0	1	0	1	0	0	0	0	0	0	0	1	0	0	0	0	0	300	-	
s_{22}	0	0	1	0	1	0	0	0	0	0	0	0	1	0	0	0	0	700	700	pivot row
	Δ Zj	-10.9 M	-14 M	-5,9 M	-7,4 M	-2 M	-2 M	1 M	1 M	1 M	1 M									
			pivot column																	

Hence, in the next tableau x_2 replaces s_{22} in the basis. The pivot element is marked grey. By pivoting and computing ΔZ_j, b and Θ, tableau 2 results from tableau 1.

On this page it is shown, how tableau 2 results from tableau 1 by pivoting.

After the replacement of the basis, the following variables are in the basis of tableau 2: h_{11}, h_{12}, h_{13}, h_{14}, s_{21} and x_2 (instead of s_{22}).

The pivot row is the last row, as marked in tableau 1. The pivot element is 1, thus pivoting is eased. For the following element E, of the second tableau the calculation will be demonstrated.

		x_1	x_2	x_4	b
Eh_{11new}	$= Eh_{11old} - 3{,}754 \cdot E_{pivot\ row}$	= 4,538 - 3,754 · 0 = 4,538	= 3,754 - 3,754 · 1 = 0	= 1,155 - 3,754 · 1 = -2,599	= 50.000 - 3,754 · 700 = 47.372
Eh_{12new}	$= Eh_{12old} - 6{,}338 \cdot E_{pivot\ row}$	= 3,754 - 6,338 · 0 = 3,754	= 6,338 - 6.338 · 1 = 0	= 2,535 - 6,338 · 1 = -3,803	= 47.000 - 6,338 · 700 = 42.264
Eh_{13new}	$= Eh_{13old} - 1{,}365 \cdot E_{pivot\ row}$	= 1,444 - 1,365 · 0 = 1,444	= 1,365 - 1,365 · 1 = 0	= 1,260 - 1,365 · 1 = -0,105	= 48.000 - 1,365 · 700 = 47.045
Eh_{14new}	$= Eh_{14old} - 2{,}535 \cdot E_{pivot\ row}$	= 1,155 - 2,535 · 0 = 1,155	= 2,535 - 2,535 · 1 = 0	= 2,400 - 2,535 · 1 = -0,135	= 43.000 - 2,535 · 700 = 41.226
Es_{21new}	$= Es_{21old} - 0 \cdot E_{pivot\ row}$	= 1 - 0 · 0 = 1	= 0 - 0 · 0 = 0	= 0 - 0 · 0 = 0	= 300 - 0 · 700 = 300
Ex_2	$= Es_{22}$ / pivot element	= 0 / 1 =0	= 1 / 1 = 1	= 1 / 1 = 1	= 700 / 1 = 700

Tableau 2

The variable x_1 has the minimum value in the row of ΔZ_j, which therefore is taken as a part of the basis of the next tableau. Dividing b by the coefficients of x_1, results in Θ, which minimum defined value exceeding zero is s_{21} .

	c_i	x_1	x_2	x_3	x_4	u_1	u_2	s_{11}	s_{12}	s_{13}	s_{14}	s_{21}	s_{22}	h_{11}	h_{12}	h_{13}	h_{14}	b	Θ	
h_{11}	- M	4,538	0	1,444	-2,599	1	0	-1	0	0	0	0	-3,754	1	0	0	0	47.372	10.440	
h_{12}	- M	3,754	0	1,365	-3,803	0	1	0	-1	0	0	0	-6,338	0	1	0	0	42.564	11.339	
h_{13}	- M	1,444	0	1,838	-0,105	1	0	0	0	-1	0	0	-1,365	0	0	1	0	47.045	32.585	
h_{14}	- M	1,155	0	1,26	-0,135	0	1	0	0	0	-1	0	-2,535	0	0	0	1	41.226	35.693	
s_{21}	0	1	0	1	0	0	0	0	0	0	0	1	0	0	0	0	0	300	300	pivot row
x_2	0	0	1	0	1	0	0	0	0	0	0	0	1	0	0	0	0	700	-	
	ΔZ_j	-10,9 M		-5.9 M	6.6 M	-2 M	-2 M	1 M	1 M	1 M	1 M		14 M							
		pivot column																		

Thus x_1 replaces s_{21} as a part of the basis of the third tableau. After replacing in the basis, tableau 2 is pivoted and b is computed. From the values of the new pivot column, the criteria for elimination Θ is calculated again. Thus tableau 3 results:

Tableau 3

	c_i	x_1	x_2	x_3	x_4	u_1	u_2	s_{11}	s_{12}	s_{13}	s_{14}	s_{21}	s_{22}	h_{11}	h_{12}	h_{13}	h_{14}	b	Θ	
h_{11}	- M	0	0	-3,093	-2,599	1	0	-1	0	0	0	-4,538	-3,754	1	0	0	0	46.011	46.011	pivot row
h_{12}	- M	0	0	-2,389	-3,803	0	1	0	-1	0	0	-3,754	-6,338	0	1	0	0	41.438	-	
h_{13}	- M	0	0	0,393	-0,105	1	0	0	0	-1	0	-1,444	-1,365	0	0	1	0	46.611	46.611	
h_{14}	- M	0	0	0,105	-0,135	0	1	0	0	0	-1	-1,155	-2,535	0	0	0	1	40.879	-	
x_1	0	1	0	1	0	0	0	0	0	0	0	1	0	0	0	0	0	300	-	
x_2	0	0	1	0	1	0	0	0	0	0	0	0	1	0	0	0	0	700	-	
	Δ Zj			5 M	6,6 M	-2 M	-2 M	1 M	1 M	1 M	1 M	1,1 M	1,4 M							
						pivot column														

In this tableau, it occurs that two different variables, namely u_1 and u_2, have the same value for ΔZ_j. This means that the value of the objective function increases equally from this corner by selecting either variable. Both variables can be selected and would give the same results. But we can not predict the number of corners of the polyhedron that must be passed either-way to reach the optimum solution. As the new basis variable we choose u_1 to replace h_{11}.

Tableau 4

In this tableau the basis variable that should be selected is well-defined. After calculation of Θ, the complete tableau 4 results.

	c_i	x_1	x_2	x_3	x_4	u_1	u_2	s_{11}	s_{12}	s_{13}	s_{14}	s_{21}	s_{22}	h_{11}	h_{12}	h_{13}	h_{14}	b	Θ	
u_1	0	0	0	-3,094	-2,599	1	0	-1	0	0	0	-4,538	-3,754	1	0	0	0	46.011	-	
h_{12}	- M	0	0	-2,389	-3,803	0	1	0	-1	0	0	-3,754	-6,338	0	1	0	0	41.438	41.438	
h_{13}	- M	0	0	3,488	2,494	0	0	1	0	-1	0	3,094	2,389	-1	0	1	0	600	-	
h_{14}	- M	0	0	0,105	-0,135	0	1	0	0	0	-1	-1,155	-2,535	0	0	0	1	40.879	40.879	pivot row
x_1	0	1	0	1	0	0	0	0	0	0	0	1	0	0	0	0	0	300	-	
x_2	0	0	1	0	1	0	0	0	0	0	0	0	1	0	0	0	0	700	-	
	Δ Zj			-1,2 M	1,4 M		-2 M	-1 M	1 M	1 M	1 M	1,8 M	6,5 M	1 M						
							pivot column													

Next u_2 replaces h_{14} as a part of the basis.

Tableau 5

As in tableau 3, two different variables, namely s_{14} and s_{11} have the same value of ΔZ_j. However, the selection of the new basis variable is not indifferent now, because the replacement of s_{11} by h_{12} breaks the Kuhn-Tucker condition that $L_x \cdot x_0 = 0$. Because x_1 is part of the solution, s_{11} may not be selected to enter the basis at the same time (Zimmermann, 1992, p. 159). Thus the new basis variable is s_{14} and h_{12} is the minimum feasible value of Θ.

	c_j	x_1	x_2	x_3	x_4	u_1	u_2	s_{11}	s_{12}	s_{13}	s_{14}	s_{21}	s_{22}	h_{11}	h_{12}	h_{13}	h_{14}	b	Θ	
u_1	0	0	0	-3,094	-2,599	1	0	-1	0	0	0	-4,538	-3,754	1	0	0	0	46.011	-	
h_{12}	- M	0	0	-2,494	-3,668	0	0	0	-1	0	1	-2,599	-3,803	0	1	0	-1	559	559	pivot row
h_{13}	- M	0	0	3,488	2,494	0	0	1	0	-1	0	3,094	2,389	-1	0	1	0	600	-	
u_2	0	0	0	0,105	-0,135	0	1	0	0	0	-1	-1,155	-2,535	0	0	0	1	40.879	40.879	
x_1	0	1	0	1	0	0	0	0	0	0	0	1	0	0	0	0	0	300	-	
x_2	0	0	1	0	1	0	0	0	0	0	0	0	1	0	0	0	0	700	-	
	Δ Zj			-0,9 M	1,2 M			-1 M	1 M	1 M	-1 M	-0,5 M	1,4 M	1 M			1 M			
											pivot column									

Tableau 6

From the selection of the pivot column the new basis variable x_3 results. Because s_{13} is no longer in the basis, the replacement in the basis can be carried out as usual. The last auxiliary variable h_{13} leaves the basis.

	c_i	x_1	x_2	x_3	x_4	u_1	u_2	s_{11}	s_{12}	s_{13}	s_{14}	s_{21}	s_{22}	h_{11}	h_{12}	h_{13}	h_{14}	b	Θ	
u_1	0	0	0	-3,094	-2,599	1	0	-1	0	0	0	-4,538	-3,754	1	0	0	0	46.011	-	
s_{14}	0	0	0	-2,494	-3,668	0	0	0	-1	0	1	-2,599	-3,803	0	1	0	-1	559	-	
h_{13}	- M	0	0	3,488	2,494	0	0	1	0	-1	0	3,094	2,389	-1	0	1	0	600	172	pivot row
u_2	0	0	0	-2,389	-3,803	0	1	0	-1	0	0	-3,754	-6,338	0	1	0	0	41.438	-	
x_1	0	1	0	1	0	0	0	0	0	0	0	1	0	0	0	0	0	300	300	
x_2	0	0	1	0	1	0	0	0	0	0	0	0	1	0	0	0	0	700	-	
	Δ Zj			-3,5 M	-2,5 M			-1 M	0	1 M		-3.1 M	-2.4 M	1 M	0		0			
				pivot column																

Tableau 7

Tableau 7 presents the optimum solution of the objective function subject to the constrants, because no ΔZ_j less than zero exists (Zimmermann, 1992, p. 72). The optimum area of the four stands with different harvesting age and risk of storm damage can be read off the tableau. The maximum expected utility results from choosing x_1 **= 128,** x_2 = **700,** x_3 = **172** and x_4 = **0**. Because x_4 does not belong to the basis it is still predetermined zero.

	c_i	x_1	x_2	x_3	x_4	u_1	u_2	s_{11}	s_{12}	s_{13}	s_{14}	s_{21}	s_{22}	h_{11}	h_{12}	h_{13}	h_{14}	b	Θ
u_1	0	0	0	0	-0,387	1	0	-0,113	0	-0,887	0	-1,793	-1,635	0,113	0	0,887	0	46.544	
s_{14}	0	0	0	0	-1,884	0	0	0,715	-1	-0,715	1	-0,387	-2,094	-0,715	1	0,715	-1	988	
x_3	0	0	0	1	0,715	0	0	0,287	0	-0,287	0	0,887	0,685	-0,287	0	0,287	0	**172**	
u_2	0	0	0	0	-2,094	0	1	0,685	-1	-0,685	0	-1,635	-4,701	-0,685	1	0,685	0	41.849	
x_1	0	1	0	0	-0,715	0	0	-0,287	0	0,287	0	0,113	-0,685	0,287	0	-0,287	0	**128**	
x_2	0	0	1	0	1	0	0	0	0	0	0	0	1	0	0	0	0	**700**	
	Δ Zj	0	0	0	0	0	0	0	0	0	0	0	0	0	0	0	0		

Managing Forest Ecosystems

1. K. von Gadow, T. Pukkala and M. Tomé (eds.): *Sustainable Forest Management.* 2000 ISBN 0-7923-6356-6
2. K. von Gadow (ed.): *Risk Analysis in Forest Management.* 2001 ISBN 0-7923-6900-9

Kluwer Academic Publishers – Dordrecht / Boston / London

MIX
Papier aus verantwortungsvollen Quellen
Paper from responsible sources
FSC® C105338

If you have any concerns about our products,
you can contact us on
ProductSafety@springernature.com

In case Publisher is established outside the EU,
the EU authorized representative is:
Springer Nature Customer Service Center GmbH
Europaplatz 3, 69115 Heidelberg, Germany

Printed by Libri Plureos GmbH
in Hamburg, Germany